ANTIBALLISTIC MISSILE DEFENCE
IN THE 1980S

Antiballistic Missile Defence in the 1980s

Edited by
Ian Bellany
University of Lancaster

and

Coit D. Blacker
Stanford University

FRANK CASS

First published 1983 in Great Britain by
FRANK CASS AND COMPANY LIMITED
Gainsborough House, 11 Gainsborough Road,
London, E11 1RS, England

and in the United States of America by
FRANK CASS AND COMPANY LIMITED
c/o Biblio Distribution Centre
81 Adams Drive, P.O. Box 327, Totowa, N.J. 07511

British Library Cataloguing in Publication Data

Antiballistic missile defence in the 1980s.—(Arms control
ISSN 0144-0381. Vol. 3; no. 2)
1. Antimissile missiles—History
I. Bellany, Ian   II. Blacker, Coit D.
III. Series
355.03′08   UG1312.A6

ISBN 0-7146-3207-4

This group of studies first appeared in a Special Issue on
'Antiballistic Missile Defence in the 1980s' of *Arms
Control*, Vol. 3, No. 2, published by Frank Cass & Co.
Ltd.

*Printed in Great Britain by
T. J. Press (Padstow) Ltd., Padstow*

Contents

Notes on Contributors — vii

Introduction — ix

Ballistic Missile Defence as a Practicable
    Proposition — *H. O. Newsham* — 1

Critical Decision Areas in the BMD Arms
    Control Debate — *Jack L. Kangas* — 12

Arms Control Implications of Ballistic
    Missile Defense — *Robert C. Gray* — 29

The ABM and American Domestic
    Politics — *Phil Williams* and *Stephen Kirby* — 46

Europe and the ABM Revival — *Lawrence Freedman* — 73

Extended Deterrence and the Protection
    of the United States ICBM Force — *Ian Bellany* — 86

Notes on Contributors

Ian Bellany is Professor and Director of CSACIS at the University of Lancaster. The author of *Australia in the Nuclear Age* and a large number of articles on defence and disarmament issues, he is currently engaged on theoretical studies in arms control.

Coit D. Blacker, the American editor of *Arms Control* is Associate Director of the Arms Control and Disarmament Program at Stanford University, where he also teaches in the Political Science Department. A Council on Foreign Relations International Affairs Fellow during academic year 1981/82, Dr Blacker has just completed a twelve month tour of duty in the office of Senator Gary Hart, where he served as special assistant for national security issues.

Lawrence Freedman is Professor and Head of the Department of War Studies at King's College, London. Prior to April 1982 he was Head of Policy Studies at the Royal Institute of International Affairs. In addition to many articles on defence and foreign policy Professor Freedman is the author of *U.S. Intelligence and the Soviet Strategic Threat*, *Britain and Nuclear Weapons* and *The Evolution of Nuclear Strategy*.

Robert C. Gray is Associate Professor of Government at Franklin and Marshall College. In 1979-80 he was a Council on Foreign Relations Fellow, serving in Washington as a policy analyst in the office of the Secretary of Defense. He is currently conducting research on U.S. nuclear strategy.

Jack Kangas studied defence policy at M.I.T. and Stanford University. He was with the office of the Secretary of Defense from 1971 to 1978 and is now a consultant with the Washington Defense Research Group Inc.

Stephen Kirby lectures in Politics at the Unversity of Hull and during academic year 1982/83 is visiting professor at the University of Tours. He has published articles on British and American defence policies and defence budgeting and is co-author (with Andrew Cox) of *Defence Budgeting in Britain and America* to be published in early 1983.

H. O. Newsham is a pseudonym.

Phil Williams is a lecturer in International Relations in the Department of Politics, University of Southampton. He is currently a Research Fellow at the Royal Institute of International Affairs where he is engaged on a project in Atlantic relations. He is author of *Crisis Management*, a co-author of *Contemporary Strategy*, and is currently completing for publication a book entitled *The Senate and U.S. Troops in Europe.*

Introduction

Just over ten years ago, in Moscow, the United States and the Soviet Union signed what has come to be regarded as the most important and enduring arms control agreement ever concluded by the two superpowers – the anti-ballistic missile treaty of May 1972. The treaty is of unlimited duration, although subject to formal review by the signatories every five years. The first such review, in 1977, passed without significant incident. The second review, which has yet to take place, promises to be rather more eventful.

In contrast to its immediate predecessor, the Reagan administration has made clear its determination to explore, carefully and aggressively, the technical possibilities for, as well as the political and military implications of, a U.S. ballistic missile defense system (BMD). Indicative of its strong interest in this regard, the administration requested over $700 million for BMD-related research and development in its fiscal year 1983 budget. The Congress, unconvinced by the logic of the administration's case and eager to find ways to trim the defense budget, reduced the Reagan request this summer by almost one half, authorizing roughly $100 million more than had been appropriated the year before. The administration, which seems not to have anticipated the hostile congressional response, has already signalled its intention to submit a new BMD funding request during the FY 1984 budget cycle that could exceed $ one billion. In short, the debate within the American policymaking community over the advisability of accelerating BMD programs, far from being over, is, in reality, about to begin. Moreover, the administration's stake in the outcome of the debate is quite high – it is extremely difficult to imagine any militarily viable permanent basing plan for the troubled MX missile that will not require at least a limited anti-ballistic (ABM) deployment.

As suggested by this analysis, the publication of this special edition of *Arms Control* could hardly be more timely. These six essays, each written by an informed and respected analyst in the field of strategic studies and arms control, represent a conscious effort to break new ground in a highly controversial and complicated issue area. Many of the essays have been in preparation since the early months of 1981 – all are the products of careful scholarship and thoughtful reflection. Among the issues discussed in detail are the development of ABM related technologies since the signing of the 1972 treaty: the domestic politics of ABM in the United States; the arms control implications of a renewed superpower competition in BMD technologies; the link between U.S. and Soviet BMD efforts and the independent nuclear forces of the United Kingdom and France; and the effect of BMD on 'extended deterrence'.

Our audience is both the policy maker who must consider and digest the interaction of technical, political, and military issues with respect to ABM, and the informed general reader, for whom little has been written on this subject over the last decade.

Above all, the editors hope that this special volume will both stimulate discussion and contribute in a positive way to the emerging debate on ballistic missile defenses. As one of the contributors notes in his introduction, the question of BMD, frequently overlooked in the arms control and strategic literature of the 1970s, could prove to be, for the United States and its allies, the most important defense issue of the 1980s.

C. D. Blacker

Ballistic Missile Defence
as a Practicable Proposition

H. O. Newsham

There is a small category of problems facing military planners that are at first glance intractable – in as much as mechanically devoting more physical resources to the problem causes the solution to recede rather than come closer – but which on closer inspection succumb to a 'software' approach rather than a 'hardware' approach. The outstanding example – which was met with in both world wars of this century – was the problem (facing the British Admiralty chiefly) of how to protect merchant shipping against submarine attack. Building new merchant ships to replace those lost was simply unfeasible since the sinking rate substantially exceeded the maximum achievable building rate. Hardware solutions such as better methods of detecting submarines, particularly the use of aerial surveillance and better methods of sinking them, by variable settings on bomb and depth charge fuses, played a useful part in shifting the odds against the submarine but the decisive innovation that turned the tables was not a new weapon but a novel (though not entirely new) and, for many, counter-intuitive way of using existing weapons – the convoy system.

A problem that has faced military planners for 25 years that also at first glance seems intractable but which may eventually succumb to a software solution is that of ballistic missile defence. The physical solution to the problem, within the guided interceptor missile technology paradigm,[1] of putting increased numbers of interceptor missiles at the disposal of the defence, is no solution at all since the attacker can far more cheaply, as a rule, add a fresh, reliable warhead to the attack than the defender can compensate by bringing an additional, reliable interceptor into play. Since the invention of intercontinental ballistic missiles (ICBMs), in the years between 1955 and 1960, and the first stirrings of interest in how to intercept them, which date from the same period, the crude physical advantage (quantifiable in the sort of cost terms employed above) has continued to move, more or less steadily, in favour of the attacker. The attacker has benefited in particular (as it was very much intended that he should) from the MIRV technique which permits the attacker to add new warheads without always having to go to the expense

1

of providing additional launch vehicles.

By comparison progress in BMD technology has been impeded, as again was fully intended, by the operation of the 1972 Moscow Agreement limiting anti-ballistic missile systems (part of SALT I), on the coming into force of which the annual rate of United States' spending on BMD research and development dropped in a step-shaped curve from approximately $600 million in 1972 to $300 million in 1975 (both figures in 1980 dollars).[2]

Whilst, as will be discussed further below, this has not prevented the appearance of some BMD hardware innovations in the period since 1972, the indirect and presumably unintended result of the financial disciplines imposed on BMD research in the United States has been to encourage a more realistic and software sensitive approach than formerly, both among those responsible for proposing new BMD schemes (the United States Army, mainly) and those to whom it falls to appraise Army proposals.

The Search for Leverage

Software solutions to the BMD problem amount essentially to ways of economising on the use of interceptors. Any arrangement providing such economies is said to provide 'leverage'.

The simplest form of leverage comes from employing interceptors with long range, designed to intercept attacking warheads above the earth's atmosphere, perhaps 10 minutes before they are due over their targets. Leverage comes from the defender deciding to offer protection only to a proportion of the targets (which in theory could be either cities, ICBM sites or bomber bases), but since the attacker does not know which targets will be preferentially defended, in order to re-achieve completely the level of destruction he originally intended, he has to assume every target is defended and reinforce his attack accordingly.

To give an illustration, suppose 100 (perfect) long range interceptors were used to defend the 1000 existing United States Minuteman sites, and let us suppose in the undefended case, the attacker has to assign only one (also perfect) warhead to each silo to be sure of destroying it. In the undefended case, then, to be sure of destroying every silo the attacker needs 1000 warheads. In the defended case, to be sure of destroying *every* silo, the attacker needs 2000 warheads. In other words 100 interceptors force the attacker to acquire 1000 additional warheads, giving a leverage of 10:1. But if the attacker could tell in advance which of the silos were being defended – which would be all too easy if the BMD interceptors had short range (they could defend only those silos within reach), or if the silos were obviously unequally important to the defender (some containing old ICBMs, say, others containing newer and more powerful models) – he could overcome the defence completely by adding an extra warhead only against the defended targets, and in this case 100 interceptors could be overcome by only 100 additional warheads.

The above simplified example illustrates quite well the tendency in ballistic missile defence schemes for leverage to depend strongly upon how much defence is attempted. If the defender wanted, say, to protect 200 of his 1000 ICBM silos, his leverage would drop to 5:1. If, as seems destined to remain the case for the foreseeable future, the cost of putting one reliable extra warhead into the attack is always less than the cost of putting one additional reliable interceptor into the defence, the point at which the balance of costs (taking leverage into account) favours neither attacker nor defender will always be struck well below complete BMD coverage of the targets in question.

The loss of confidence in the BMD concepts embodied in the U.S. Army's Safeguard system of long range interceptors after the signing of the 1972 Moscow Agreement, and the Army's subsequent transfer of affections to its new long range Overlay system, which is discussed in detail below, suggests very strongly that even leverages of the order of 10 to 1 were not sufficient to offset the drawbacks inherent in Safeguard and still extend BMD cover to a worthwhile proportion of the United States' ICBM force. The principal drawbacks were three in number: two stemming from the need to rely upon radar to identify incoming warheads and steer the interceptors, and one stemming from the so-called decoy problem.

To detect and steer at long ranges (3000 kilometres or more), large, powerful, and hence very costly, ground based radars are necessary. These radars have therefore themselves to be defended against selective attack, and it is difficult to see how this can be done at acceptable cost (a second line of defence based on short range interceptors to guard the radars is – as we shall see – very vulnerable to saturation attack). The second drawback arises from the possibility that the attacker will deliberately detonate 'pre-cursor' warheads high in the atmosphere to create extensive zones of ionisation through which radar beams probing from ground level will find it difficult to penetrate and behind which the main body of the attack can be 'hidden'.

The decoy problem arises because above the atmosphere nothing in the flight path of a menacing object detected by radar need signal whether it is a real re-entry vehicle carrying a nuclear warhead and weighing upwards of a few hundred kilograms, or a cheap aluminized plastic balloon designed to give the same image on a radar screen but weighing at most only a few kilograms. For the sacrifice of one real re-entry vehicle the attacker can put at least ten decoys into the attack (some of the weight difference will be taken up by the mechanism used to dispense

the decoys in a properly convincing manner). Thus a maximum hypothetical attack of 8000 warheads can be made to look like an attack of
17,000 warheads (7000 + 10 × 1000). If there was in fact no way of
discriminating between decoys and real re-entry vehicles, in this
particular example the defender would inevitably waste more than half
of his interceptors firing at non-targets.[3]

The U.S. Army's new Overlay system, which it must be stressed is still
at the drawing board stage, tackles the first two of Safeguard's main
weaknesses but not the third.[4] Radar is employed only as one of several
technically and locationally distinct means available to give warning of
an attack. In Overlay the position, strength and heading of the attack are
sensed by infra-red detection probes despatched above the atmosphere
by rocket as soon as warning of attack is received. This information is
used to guide interceptor missiles towards the battle zone. The interceptors are themselves fitted with multiple independent non-nuclear
warheads, which on release from the interceptor missile home to the
point of impact, guided by their own individual infra-red sensors.

In addition to the valuable but somewhat negative feature of being far
less dependent than Safeguard on radar, Overlay offers two additional
features of its own. First, multiple warheads promise to keep the system
cost for each interceptor warhead reasonably low – in Safeguard each
interceptor warhead expensively required its own launcher. Secondly the
use of non-nuclear warheads very much reduces the significance of an
accidental firing of an Overlay interceptor and moreover makes the
delegation of authority to fire (technically always desirable whenever
time is short) politically very much simpler.

The flaw in the current Overlay plan is that it is no answer to the decoy
problem. Decoys which appear to an infra-red sensor no different from
warheads can probably be constructed as cheaply and convincingly and
at as little cost to launcher weight carrying capacity as decoys designed to
deceive radar. The way ahead for Overlay may lie through advances in
microcircuitry making it possible to equip the detection probe and the
interceptor's multiple warheads with sensors that can 'see' at other
wavelengths in addition to the infra-red. Decoys that 'looked' the same
as real warheads both to infra-red and to, let us say, as a completely
hypothetical example, millimetric radar might actually be difficult for
the attacker to build without incurring substantial penalties in cost or in
weight terms.[5]

As its name implies Overlay is not designed to work alone. In a general
sense, quite independent of the actual hardware techniques employed,
layering defence – having at least two lines of defence, in other
words – which means using a second line or underlay specifically to intercept attacking warheads that penetrate the first line, does by itself

invariably create leverage. It is simply more efficient (we are now talking of real, i.e. less than perfect, interceptors) as a general rule to wait to correct for failures in intercepting an attack until after it is known which of the first wave of intercepts have failed, than to try in a statistical way to anticipate and cover for failures by putting every interceptor in the first line and allocating two or more interceptor warheads to each attacking warhead.[6]

On the other hand were the second layer to rely upon short-range interceptors, the leverage from layering as such would still be there, but it then becomes impossible to continue to extract the full leverage benefits of preferential defence. Partly to continue to keep the attacker in the dark as to which targets (these may be cities or ICBM silos) will be preferentially defended, any short-range interceptor coverage must be given to all targets of a particular category, or to none. If it is given to all it will function as designed – as a second line of defence – only in a proportion of cases (over targets to which upper layer coverage had been preferentially extended) and much less well than designed – perhaps wholly inadequately – in those cases where it is the only line of defence between the target and the full force of the attack.

The leverage that comes with preferential defence and long range interceptors was important to the U.S. Army's Sentinel BMD system, which in the late 1960s was put forward by the Johnson Administration as a means of protecting United States' cities against the projected threat of a light and primitive Chinese ICBM attack. Essentially the same system design and technology were co-opted by the Nixon Administration under the new name of Safeguard, but re-assigned to the job of protecting United States' ICBM silos against Soviet attack. As we have seen Safeguard was too dependent on a handful of high powered radars and too easily deceived by decoys to have been much use against such a sophisticated attacker.

Whilst Safeguard was somewhat belatedly discovered to be not up to the job, technically, and the Sentinel plan became politically redundant, Sentinel's paper predecessor Nike-X which envisaged long range and short range interceptors in a full blown two tier system fell victim to second thoughts about what happens to cost effectiveness when leverage through preferential defence becomes diluted by the existence of a second line of defence based on short range non-preferential interceptors.[7]

What survived from the Nike-X scheme to play a small part in both Sentinel and Safeguard, although these were essentially area defence schemes relying on long range interceptors, was the idea of the short range or point defence interceptor. The powerful attraction of the short range interceptor is that it can solve the decoy problem. And the chief hardware innovations of the past decade in BMD have probably been the transformations made to the Sprint short range interceptor of Nike-X and Sentinel, which by reducing its ceiling of operation from 30,000 metres to 15,000 metres not only allows the missile and its supporting radar to be much more compact than before but also allows the filtering effect of the atmosphere more time to separate decoys from real warheads.

By postponing interception to such low altitudes – until the attacking warheads are within about 5 seconds of impact – computer assisted radar tracking can make use of atmospheric drag essentially to permit a remote but very accurate measurement to be made of the *weight* of the menacing object, and the heaviest objects must be the warheads.[8] Since the radars can be physically small, they are also relatively cheap and easy to protect by passive measures. Furthermore the interceptors can be given nuclear warheads of the enhanced radiation type which have a large radius of destruction (a kilometre or more), with a kill mechanism (neutrons) against which it is virtually impossible to harden attacking warheads (without incurring huge weight penalties), and are set to explode at altitudes where the adverse effects of nuclear bursts on radar propagation are at a minimum.

The price to be paid is that delaying interception until so late makes it relatively easy for the attacker to saturate the point defence by directing a parcel of warheads at it in such close order as to leave the defender no time normally to make more than one intercept. But with point defence as a second layer, behind a long range interceptor outer layer of the Overlay type, Overlay interceptors will tend to interfere with the attacker's attempts to bunch his warheads into the kind of packets to which the point defence is most vulnerable. Synergistic factors of this type are also an important source of leverage.

Lest it should seem from the foregoing that leverage is the magic key to ballistic missile defence that opens all doors, it is important not to overlook some of the limitations of BMD.

Even if the decoy problem could be solved, preferential defence based on long range interceptors with the goal of protecting no more than a sizeable minority of targets would require a public much more somnolent than those of recent years before it even began to sound like a politically attractive means of defending American cities.[9] Low altitude defence, even with Overlay or something similar in support to break up the kind of attacks point defence is most vulnerable to, is not at its best in the defence of cities either, since the attacker can pretty well achieve his objectives by exploding high yield warheads at altitudes well above the operational ceiling of the point defence system.[10] Whether these short-comings are the cause, or in some way a consequence, of the decline over the past fifteen years in the amount of official interest shown in the United States in the city defence question is a moot point.

Low altitude defence of the existing Minuteman ICBM silos, without an Overlay, simply has no leverage at all and the attacker would need to add only one or two warheads to the two that would normally be enough to destroy an undefended silo, totally to nullify the defender's expenditure on BMD. Even with an Overlay in support, low altitude defence of Minuteman would dilute the leverage derived from preferential defence. What is more, the single shot-only capability of low altitude defence places stern demands on the efficiency of supporting long range interceptors. Out of an eight warhead attack on a single silo, more often than not two would leak through an engagement by eight Overlay interceptors (and saturate the low altitude defence) unless the interceptors could maintain an individual success rate in excess of 80 per cent.[11]

And yet, the perceived vulnerability of the Minuteman silos to attack, and the proposal by the Carter Administration to secure the survival of a portion of the United States' ICBM force by deceptively basing one MX missile in each of 200 clusters of lightly hardened multiple protective shelters (MPS), 23 shelters to a cluster, in Utah and Nevada, opened up a prospect of a source of leverage for BMD that was to prove almost irresistibly tempting.

In essence the MPS basing plan relied upon the attacker having no means of knowing which of the shelters in a cluster held the MX missile at any time. To be sure of hitting one MX the attacker would need to assign the equivalent of 23 perfectly reliable warheads. But 4600 separate aim points, which would have cost in the region of $46,000 million[12] were rather few, it was argued, in the context of a Soviet ICBM force equipped with 8200 warheads (the maximum MIRV-launched number allowed under the unratified SALT II treaty signed in Vienna in 1979).

The United States Army proposed a modification to the MPS scheme wherein each 23-shelter cluster would hide not only an MX missile but also one short range low altitude BMD interceptor of the kind described above, complete with its compact radar and data processing package (the so-called LoADS – low altitude defence system).[13] Since the LoADS would need to intercept only the particular attacking warhead actually heading for the shelter containing the MX missile, in order to be sure of destroying the MX, in the presence of LoADS the attacker would need to fire 46 warheads at each cluster, giving the defender a leverage of 23 to 1. The cost of the LoADS at 1980 prices was estimated to lie between $6 and $9 thousand million,[14] adding roughly 20 per cent to the cost of an undefended shelter.

LoADS was not enough to save the MPS scheme, but it appears to have come close to doing so. Whilst political arguments – some connected with the adverse environmental impact of MPS and others to do with the fact that MPS was anyway the brainchild of a previous Administration – played a part, the technological case for the MPS/LoADS combination was good but not copper bottomed. Because of the saturation effect there would have been no point, normally, in equipping the LoADS unit in each cluster with more than one interceptor and a Soviet attack capability in excess of 9200 warheads could not then have been met by an incremental improvement to LoADS any more than (as its opponents were quick to realise) the bare MPS could have been incrementally improved to cope with a Soviet attack capability going beyond 4600 warheads.[15] Authoritative doubts were also expressed as to whether LoADS was quite as capable as the U.S. Army was making it out to be. The Comptroller General's Office was put in mind of the claims that the Army had made in the late 1960s for Safeguard and noted acidly that experience had shown how far in excess of actual performance these had been pitched.[16] The Office of Technology Assessment was less sceptical but pointed to two specific potential flaws. First it was not clear to them that the compact LoADS radar and data processing package, integral with the interceptor launcher, once it had emerged from its protective shelter to grapple with an attack on its cluster, would be robust enough to withstand the effects of nuclear explosions taking place as a result of attacks on other, possibly empty, shelters in the cluster.[17] Nor was it clear what redress would be available to the defender if the attacker found a way of 'smoking out' the LoADS interceptor by, for instance, firing a barrage at a cluster, noting through some kind of yet to be developed, but theoretically possible, reconnaissance arrangement which shelter within the cluster had received LoADS cover, and then aiming a fresh attacking warhead or two at that shelter alone.[18]

It remains probable that the eventual means chosen of basing the MX missile, if there is indeed to be one, will feature BMD in some role. Since Overlay technology is not going to be available for several years at the soonest and MX deployment is presumably a matter of urgency, options within the paradigm narrow to LoADS or something very similar.

The contending plans for an MX land basing scheme to replace the MPS that have received a public airing do not seem immediately to be natural candidates for supplementary ballistic missile defence. The Closely-Spaced Basing plan – which envisages 100 MX missiles emplaced in very hard silos about 500 metres apart and which relies upon the first attacking warhead to act, in effect, as a large BMD warhead as far as the other attacking warheads in the same wave are concerned and on the debris thrown up by the first attacking warhead to act as an abrasive shield which will destroy the rapidly descending warheads in subsequent waves, but which need not prevent the launch out of MX missiles in gaps between one wave and the next – supposedly includes a BMD provision.[19] But there is no practical point in BMD without leverage and it is very hard in this case to see where leverage is going to come from, given the saturation problem.

On the other hand Safeguard quite certainly, and MPS/LoADS arguably, were instances where BMD was applied as an afterthought to patch up a passive defensive basing scheme for ICBMs that had not been originally designed with BMD in mind.[20] It is possible that a scheme designed from the beginning to take advantage of the properties of LoADS will prove to be more workable.

Consequences for Arms Control

Whether or not a particular BMD scheme to defend ICBM sites falls within the quantitative and qualitative limits set by the 1972 Moscow Agreement is an important question, but an obvious one that nobody is going to overlook. What may be less obvious is the demonstration of the inability of an arms control agreement which perhaps of necessity concentrates on limiting the size and/or technical characteristics of a particular class of weapon to do much about the phenomenon of leverage. Arms control may limit hardware and hardware innovations but software innovations that substantially increase the military effectiveness of a limited number of units may as a result even be encouraged.

In principle leverage through software innovation can occur in any class of weaponry. It is fortuitous that the convoy example and the MPS/LoADS example both involved defensive measures (and anyway the theoretical leverage to be got from long range interceptors used preferentially in city defence is not, of course, unambiguously defensive in terms of a balance of deterrence). And equally there are many roads to leverage; it happens unfortunately to be the case that the getting of leverage in BMD, and counter-measures by the attacker to minimise leverage, appear to concentrate upon denying the enemy information. Safeguard got its leverage from preferential defence, through the other side's not knowing which ICBM silos would be defended: this was countered in part by the decoy, relying upon the defender not knowing which was a real target for his interceptors and which was a dummy. MPS/LoADS got its leverage essentially through denying the attacker knowledge of the whereabouts of the MX ICBM and of the LoADS unit.

The importance to defence of ignorance buttressed by secrecy is not new, the safety of the submarine borne ballistic missile force of both sides depends upon it and the MPS even without LoADS was to be built around it. But should its role be extended – as seems certain if BMD is ever deployed with serious intent – it can only decrease the area of its military activities that either side is prepared to be open about and make the gathering of knowledge about the enemy's capabilities, without a certain irreducible minimum of which arms control cannot function, more difficult than it already is.

NOTES

1. The interceptor missile paradigm may be outflanked either in the high technology direction, through directed energy weapons or space-based BMD systems, or in the low technology direction through, for instance, nuclear land-mines planted near ICBM silos and programmed to explode and throw clouds of abrasive debris into the air on receipt of warning that an attack on the silos is imminent. Neither of these alternative paradigms is taken seriously in this paper.
2. Raymond E. Starsman, *Ballistic Missile Defence and Deceptive Basing*, The National Defense University, Monograph Series No. 81-1, Washington D.C., 1981, p. 2.
3. There is, equally, nothing to stop the attacker making cosmetic alterations to his warheads to make them 'look' more like decoys.
4. This description of Overlay is taken from *MX Missile Basing*, Congress of the United States, Office of Technology Assessment, Washington D.C., September 1981, pp. 129, 139.
5. Some way of illuminating the attacking warheads and accompanying decoys with a radar beam would be necessary, in this example. Infra-red, by contrast, is a passive system which works by detecting infra-red radiation naturally emitted by the interceptor's targets.
6. In exactly the same way an attack by ICBM (MIRV) warheads on the ICBM silos of the other side could in theory be accomplished more cheaply in terms of warheads if the attacker could attack in waves, using one warhead for each silo in the first wave but in subsequent waves aiming only at those silos that had survived the preceding waves. It is probably more difficult for an attacker operating ICBMs to get hold of the information he needs for this kind of 'shoot-look-shoot' than it is for a defender operating a BMD. Leverages from layering as such depend upon how accurate the 'shooting' is, but a typical figure would be a factor of 2.
7. In fact the scope for leverage through preferential defence gets narrowed whenever any kind of non-preferential defence is employed as a second line – even super hardening for ICBM silos, or civil defence shelters for city populations.
8. If the heaviest objects were decoys, all well and good because this would mean that the load carrying capacity of the attacking missile launcher would have been largely taken up by heavy but harmless decoys, leaving the attacker much less space for warheads. The BMD interception, in this case, would have taken place in the mind of the attacker, rather than in the atmosphere.
9. But it might be more palatably presented to the public as a 'thin' defence designed to cope only with accidental or small 'demonstration' attacks.
10. On the other hand it should be stressed that in forcing the attacker to use high yield warheads, which are heavy, he will have been forced as a result to take up weight carrying capacity in his launchers that might otherwise have been allocated to large numbers of lighter, lower yield warheads, and forced therefore to reduce the number of targets the defender's BMD has to cope with, to the latter's advantage.
11. Eighty per cent interception means 20 per cent leakage and, in this example, on average 1.6 (20 per cent of 8) leakers per engagement. For a more precise guide to calculations of this kind see *MX Missile Basing*, pp. 132, 134.
12. This is an estimate, in 1980 dollars, for the total cost of a 4600 shelter system over fifteen years, including the cost of the MX missiles. The $10 million a shelter drops to nearer $8 million for an 8250 shelter model. See *MX Missile Basing*, p. 88.

13. Hiding the LoADS is essential. A separate, openly based LoADS adjacent to each cluster, say, would be no good since it would cost the attacker only two (or at most three) perfect warheads to destroy it, after which he could proceed with a 23 warhead attack on the MPS cluster.

14. *MX Missile Basing*, p. 125, gives the higher figure; Starsman, *op. cit.*, p. 28, gives the lower.

15. Estimating the future size of the Soviet 'threat' is rarely rewarding, but assuming, rather unrealistically, that a Soviet expansion of the number of warheads carried by their ICBMs will encounter no bottlenecks in plutonium production or other difficulties in super-MIRV-ing the SS-18, calculations usually proceed as follows: SALT II places a ceiling of 820 on land based ICBMs equipped with MIRVs: a limit restricting all ICBMs and SLBMs to the maximum number of warheads with which they have already been flight-tested: and a limit of one new model of ICBM, which must itself carry no more than ten warheads. If the Soviets were to do nothing other than load each of the ICBMs they had in service in 1981 with the maximum number of warheads with which each had been tested, this would give them 6080 ICBM warheads (actual 1981 total, about 5400). More realistically perhaps, if the Soviets were in addition to disband 62 of their SS-11 ICBMs (these are allowed only single warheads, the most they have been tested with) and to replace them with the same number of the much more modern SS-19 which has been tested with six warheads, this would give them a total of 820 MIRV ICBMs (actual 1981 total, about 758) and a total of 6390 ICBM warheads. It is possible that they could decide to scrap all of their MIRV ICBMs apart from the 308 'heavy' SS-18 allowed them under SALT II (and tested with ten warheads), and build 512 new 'single model' ICBMs of the 'light' type permitted under SALT II and equip each with ten MIRV warheads. This would bring the Soviet ICBM MIRV warhead total to 8200, and their total of ICBM warheads (MIRV and non MIRV) to 8778.

The reason why the Soviets are expected to spare the SS-18 from this putative modernisation programme is that it gives them a means if need be of rapidly going beyond the SALT II 8200 MIRV ICBM warhead total, without having to build either new silos or new missiles. The unusually large size of the SS-18 (it can launch seven and a half tonnes of warheads) means that if it were fitted with warheads equivalent to the current United States' Minuteman III type (165 kiloton yield and weighing under 250 kilograms each) it would carry 30 on each missile and add 6160 to the SALT II restricted Soviet MIRV ICBM warhead total of 8200. Details of the actual Soviet capability as of 1981 are taken from *The Military Balance 1981-82*, IISS, London, 1982.

16. *Potential of the LoAD Ballistic Missile Defense System for Protecting the MX Missile System*, Report by the Comptroller General of the United States, Unclassified Digest, C-PSAD-81-2, p. iii.

17. *MX Missile Basing*, pp. 122, 123.

18. *Op. cit.*, p. 125.

19. The *New York Times*, May 26, 1982, p. 6, reported that closely-spaced basing, called by it the 'cluster plan' (and also known as 'dense pack'), involved the use of interceptor missiles, without specifying how.

20. In the case of MPS it would be very difficult to be sure that no thought of its potential for stretching to include a BMD capability ever entered the heads of its designers.

Critical Decision Areas
in the BMD Arms Control Debate

Jack L. Kangas

Introduction

Like one of John Le Carré's almost forgotten agents, ballistic missile defense (BMD) is coming out of the cold after more than ten years of neglect. The evidence so far suggests that BMD will be central to the strategic policy debates of the 1980s and could prove to be the most important defense issue in the United States in this decade, rivaled in significance perhaps only by the related issue of requirements for weapon deployments in space.

The ABM Treaty signed by the United States and the Soviet Union in 1972 is an agreement in which each nation agrees not to deploy a BMD system for the defense of the territory of its country nor to build a defense for an individual region except for systems at two allowed sites. It limits each side, *inter alia*, to no more than 100 ABM launcher and interceptor missiles at each site and prohibits the development, testing, or deployment of sea-based, air-based, space-based, or mobile land-based ABM systems.

Many hailed the ABM Treaty at the time it was signed as the most significant arms control agreement of the postwar period and today there are many who believe it to be the vital centerpiece of the superpowers' arms control dialogue. Serious questions are being raised in the early 1980s, however, about the advisability of continuing to adhere to the Treaty. Some have recommended that certain provisions of the Treaty be modified to allow for less restrictive testing and deployment constraints and others have argued that the United States withdraw from the Treaty entirely.

The primary impetus for modifying or withdrawing from the ABM Treaty stems from concerns about the vulnerability of the U.S. ICBM force and expectations about the potential role of BMD in assuring the survivability of the MX missile. The ICBM modernization program of the Reagan Administration announced in the autumn of 1981 specified three long-term basing options for MX: continuous airborne patrol aircraft; deep underground basing; and ballistic missile defense. The

12

choice among the options was scheduled to be made no later than 1984. Secretary of Defense Weinberger later instructed the Air Force and the Army to work out an integrated basing plan that would take BMD into account as a long-term option for the defense of land-based ICBMs and other (unspecified) strategic assets. Defended and defended-deceptive basing approaches for MX deployment were to be developed in the plan. The initial operating capability (IOC) for BMD under the Department of Defense instructions was given as 1988, with emphasis given to the need for an accelerated development program.

More recently, there has emerged yet another basing option for MX: the so-called Dense Pack or Closely-Spaced Basing (CSB) configuration. This option would rely for its effectiveness on fratricide of the attacking Soviet reentry vehicles and the superhardening of the capsules containing the MX missiles. BMD is being considered at present in connection with CSB.

A key question U.S. policymakers might (perhaps should) ask as they explore the possible need to revise the government's arms control position on BMD is: how does the *decision environment* of the late 1960s and early 1970s (the period immediately prior to the signing of the ABM Treaty) compare with that of the early 1980s? In other words, do any changes in the past ten years or so argue for a different policy perspective on BMD? If so, what would or should that perspective be and what governmental policies does it imply? It is to these questions that the present article is addressed.

The concept of decision environment as used here is broken down for analytical purposes into five, interrelated *decision areas*: (1) the status of BMD technology, (2) definition of the threat, (3) BMD deployment policy, (4) strategic employment doctrine, and (5) strategic arms control.[1]

BMD Technology

Detailed criticism of the BMD technology envisaged for the 1980s is by and large so far difficult to find, but the criticism that does exist is reminiscent of the charges made against Sentinel and Safeguard. One critic, active in both periods, recently said:

> The technical question is whether the highly complex system required to cope with the real and massive attacks that we might expect can in fact be designed and made operational. Imagine the computer capable of discriminating reentry vehicles from cleverly designed decoys and space junk and then of passing the data to a ground-based command and control center that, in turn, manages a battle.[2]

Other critics have recently argued that terminal defenses are feasible but not cost-effective because they can be overwhelmed.[3] Such criticism is precisely of the kind made in the period prior to the signing of the ABM Treaty.

For example, in one critique of Safeguard it was pointed out that the 'problem is to meet the many rigorous and interacting requirements for an actual operational system: the apparatus itself is highly complex in all its components – missiles, radars and computers. Each separate part must function well not only by itself but in intricate coordination with the others.'[4] Concerning decoys, this critique argued that the long-range radar associated with Safeguard had limited capability to discriminate among different objects and that the enemy could fill the sky with so many objects that the computer couldn't handle the data-handling job. With respect to 'cost exchange', the argument was simply that 'Any ABM system can be overwhelmed simply by attacking the target with more offensive missiles than there are defensive missiles available to respond.'

Today's proponents of BMD technology distinguish among three different categories of BMD systems: conventional terminal or endoatmospheric defense systems; exoatmospheric defense; and exotic systems. In addition, they identify six functional requirements common to the three categories: (1) acquisition, (2) tracking, (3) bulk filtering and discrimination, (4) target kill, (5) kill assessment, and (6) communication, command and control (C^3). These proponents claim that acquisition and tracking 'can be adequately performed' by terminal and exoatmospheric systems but that bulk filtering and discrimination remain problematic and even controversial within the scientific community. With respect to target kill, nuclear interceptor warheads are said to raise no doubts about their effectiveness but non-nuclear warheads have yet to be successfully demonstrated. Kill assessment and C^3 are identified as 'non-trivial engineering challenges' but which 'do not require fundamental scientific challenges to achieve.'[5]

In general terms, proponents contend that conventional terminal defense systems 'can be confidently predicted to work in ICBM defense in the 1980s'; that exoatmospheric systems 'could be within grasp for the early 1990s, if aggressively pursued in R&D'; and that exotic systems 'promise revolutionary improvements in BMD and should be vigorously investigated in advanced development.'[6]

Supporters of BMD deployments at the Los Alamos Scientific Laboratory (LASL) claim that current technologies are quite different from predecessor technologies used in the earlier Safeguard system. The greatest drawback of Safeguard according to these scientists was the large ground-based radars: 'It was soon recognized that the first Spartan nuclear explosions would render large regions of the atmosphere opaque to radar propagation, thereby blinding the radars and making them vulnerable to attack.'[7] This limitation is said to be overcome with modern technologies. Small, high-resolution, sensitive, long wave-length infrared detectors replace large ground-based radars in the current designs for exoatmospheric defense and the radars associated with endoatmospheric defense are of the phased-array type and are much smaller than Safeguard's missile site radar, allowing for deceptive basing.

The LASL scientists also claim that the BMD technology of the 1980s represents a considerable step forward compared to ten years ago in other key areas such as computer technology, discrimination by radar signatures, and requirements for maintaining a favorable (cost-effective) ratio between interceptor warheads and incoming reentry vehicles. However, they acknowledge a number of outstanding technical issues which require further study. With respect to exoatmospheric issues, their views are condensed in the following passage:

> These issues are dominated by concern over extreme system complexity. Some analysts have considerable reservations about system operability; others are optimistic. Large-scale simulations in progress lend credence to system operability. Other issues include operability of sensors, computers, communications, and interceptors in a nuclear environment, potential for means of overcoming infrared discrimination, and integration of active defense with an already strained national command-communications-control system. The special problems of attacks launched by nuclear submarines lying close to American shores are particularly stressing to exoatmospheric defense, owing to the short flight times. Intercepting such attacks would require previous placement of threat assessment sensors or—at reduced efficiency—operation without them. In addition, a number of actions could be taken by the Soviets in response to our installation of an exoatmospheric defense system that could degrade its capabilities. They include maneuvering reentry vehicles, defense suppression attacks, and new decoy techniques.[8]

Against this background of potential problem areas the Los Alamos scientists remain committed to a program of examining future applications of exoatmospheric defenses.

With respect to endoatmospheric systems, the assessment of the Los Alamos scientists is that low-altitude defense is a 'relatively low-technical-risk system':

> However, the stressful nuclear environment envisioned requires interceptor and radar hardness values exceeding those of predecessor systems. Ability to defeat an intense attack against any

single target will be limited because, with the very short time available for acquisition, track, and intercept, multiple sequential intercepts will be difficult. The limited space available for interception would also reduce the ability of the system to cope with repeated attack, due to nuclear-fireball interference with radar propagation and to interceptor-interceptor fratricide. In such a dense attack, fratricide between attacking warheads could also be a problem for the attacker.[9]

The technical risk for endoatmospheric defense for the 1980s is nevertheless considered manageable by the Los Alamos analysts and they recommend a commitment to a terminal defense deployment.

BMD analysts at the Ballistic Missile Defense Systems Command and at the Los Alamos Scientific Laboratory have together developed in recent years a further BMD concept, called Layered Defense, which combines a terminal or endoatmospheric system and an exoatmospheric system. They identify this concept as the preferred approach to any BMD mission because leakage values are multiplicative and cost advantages over conventional BMD systems are said to be achieved through several leverage factors: adaptive preferential defense, multiple non-nuclear kill interceptors and low leakage.

From the foregoing one can perhaps generalize and say that while a number of dramatic advances have been made, and with others arguably just over the horizon in the area of BMD technology, there remain very serious questions about the viability of both endoatmospheric and exoatmospheric systems. The questions are being asked by both the critics and the supporters of BMD deployments. Critics tend to be skeptical about the ability to design systems that would function adequately in a nuclear attack environment, while supporters appear confident that the current state of BMD technology warrants a commitment to development and eventual deployment. In many respects, the technical issues being debated today are very similar to the issues debated ten years ago and many of them are likely to remain unresolved, for they involve inherent uncertainties such as the likely attack patterns to be adopted by the Soviet Union in actual strikes and how BMD systems would function under real as opposed to theoretical or simulated conditions.

Threat Definition

In the period prior to the signing of the ABM Treaty there was considerable confusion concerning the threat against which BMD deployments were supposed to be directed. Before the mid-1960s the BMD programs of the Department of Defense (Wizard and Nike-Zeus/Nike-X) were clearly directed against a Soviet ballistic missile threat, but during the Johnson Administration the focus was dramatically shifted to the possibility of a Chinese attack. In an address in San Francisco in September 1967, then Secretary of Defense Robert McNamara announced the Administration's decision to proceed with a new project, Sentinel, the basic purpose of which was to protect against a small Chinese attack and accidental or unauthorized launches of missiles, but which also was to provide some measure of protection for Minuteman silos against a Soviet ICBM threat.

An equally dramatic policy reversal took place within the next two years when, in March 1969, President Nixon announced his intention to deploy a BMD system, renamed Safeguard, that would defend the American people against a Chinese missile capability projected out to a ten year period and against the possibility of accidental launches from any source, but that would be more directly oriented to the requirement to protect U.S. land-based ICBM forces against a Soviet ballistic missile attack. Policymakers in this early period clearly were uncertain about whether the missile threat stemmed from Soviet or Chinese capabilities or from some Nth country projected in the mid-term to have a capability to target the continental United States.

In the early 1980s there is so far very limited, if any, discussion of Chinese or Nth country threats in BMD discussions, even though China's nuclear weapons programs have matured and the prospects for further proliferation of nuclear weapons appear greater today than they were a decade ago. The inattention to China can be explained largely by the U.S. policy of 'normalizing' its relations with the PRC and avoiding public discussion of a potential Chinese nuclear threat.

Policy thinking today in the BMD context is relatively 'threat specific': the threat is the Soviet ICBM force targeted on the Minuteman silos and, if and when deployed, the land-based MX missiles. There is almost universal acceptance today within the U.S. planning community of the vulnerability of the current ICBM force. At issue in the missile defense controversy of the late 1960s and early 1970s was the question of how soon Minuteman would become vulnerable to the SS-9 ICBM and to a large extent the debate centered on analysts' different assumptions about the blast resistance of the silos and the payload of the attacking missiles.[10] Today, there are only a handful of analysts who challenge the argument that the Minuteman silos are vulnerable, their basic contention being that the reliability and the accuracy of the Soviet SS-18s and SS-19s would be considerably degraded when actually fired on polar trajectories as opposed to the case where they are launched from test centers at Tyuratam or Plesetsk.[11]

In a more general sense, in the period prior to the signing of the ABM Treaty, the immediacy of the Soviet ICBM threat was not strongly felt in the policy and planning community. There was a belief that sufficient time remained to 'fix' the ICBM vulnerability problem. In the early 1980s, the policy consensus is that time has run out and the United States will be at a clear strategic disadvantage in terms of being subject to a Soviet disarming attack until a secure basing mode is found for the ICBM force.

Defining the Soviet threat takes into account Soviet BMD capabilities as well as its offensive capabilities. In the period prior to the signing of the ABM Treaty, there was considerable concern among U.S. defense planners that the Soviet Union could undertake an ambitious BMD program that might lead to large-scale deployments. Such a program was thought to be consistent with the traditional Soviet emphasis on defensive capabilities. The ABM Treaty would prevent the Soviets from deploying an effective BMD system. While the Soviets have apparently never expressed in the past ten years any official dissatisfaction with the ABM Treaty or interest in revising it, their views on the subject could change as they also (together with the United States) look for ways to cope with a changing strategic environment. In particular, the Soviet Union faces a different kind of Chinese nuclear threat in the 1980s and beyond, compared to that which existed in the early 1970s.

Recent detection of construction activity at the BMD launcher sites around Moscow has led to some concern about a possibly more serious Soviet interest in BMD deployments developing in the 1980s. Against the background of a decade of heavy investment in BMD research and development, attention among some U.S. planners is turning once again toward the BMD 'breakout' capability of the Soviet Union.

Deployment Policy

Under the Sentinel program the location of the long-range interceptor missile (Spartan) sites was to be mainly near cities. Public demonstrations against BMD deployments under Sentinel took place in New York, Chicago and other cities. The Safeguard program, however, shifted the deployment plans specifically to four Minuteman bases, seven Strategic Air Command bases, and to Washington D.C.

The decision under Safeguard to specify Washington, D.C., as a specific site appears in retrospect to have been the result of SALT considerations. The Russians were deploying their Galosh system around Moscow, and some may have believed that the only way to get further Congressional support for a U.S. BMD program was to argue that the United States shouldn't be denied the kind of capability being pursued by the Soviet Union. Even though a defense of the National Command Authority (NCA) might be less than fully effective, the deployment would, according to this line of reasoning, protect options for growth of the system. The problem, however, was that Congressional sentiment was almost uniformly opposed to defense of the national capital and Congress was prepared to allocate funds only for the purpose of defending ICBM complexes at Grand Forks and Malmstrom.

If those involved in the early BMD debate failed to make a clear and convincing case regarding the specific threat against which BMD deployments were to be directed, their failure was even greater (by their own admission) when it came to deciding where to deploy the system once it was built. In this instance, the lack of a coherent deployment policy was not the result of indecision about the threat (described above) but rather was due to requirements generated in the SALT process. When Kissinger said years later that 'the proposal for NCA was a first-class blunder; it made no substantive sense whatever,'[12] few were prepared to argue the point.

While the ABM Treaty signed in October 1972 permitted each side to deploy ABM systems in two deployment areas, i.e., an ICBM field and the site of the NCA, the number of permitted areas was subsequently reduced to one by the Protocol to the ABM Treaty signed in July 1974. The Protocol permits a one-time switch, allowing the United States to defend Washington, D.C. instead of certain of its ICBM fields and the Soviet Union to defend ICBMs instead of Moscow. In recent years there has been only limited discussion in the United States, however, about the possible advantages of defending the national capital with ballistic missile defenses. The subject is sometimes brought up in connection with discussions on maintaining continuity of government under nuclear attack.

The decision of the Reagan Administration to cancel the so-called Multiple Protective Shelter (MPS) basing plan for the MX missile has denied the Army the opportunity to couple BMD deployments to what many still consider the preferred basing mode for the MX. Current plans are to deploy the first 40 MX missiles in existing Minuteman silos (Titan silo emplacement was finally rejected on grounds that it was neither cost-effective, consistent with advanced C^3 systems, nor suitable for BMD) but allowance is made for such deployment beyond the first 40 missiles.

Current thinking within the Department of Defense appears to be focused on the prospect of using the Grand Forks site to defend the MX missiles in the CSB mode. Recognition is given to the need to increase the effectiveness of the Grand Forks facility to do this job, by modernizing ABM radars, data processing equipment, and new long-range interceptors. Such measures could be implemented within the constraints of the ABM Treaty. At some point, however, an effective ballistic missile defense of MX in the CSB mode is expected to require in all likelihood collocation of the defenses with the offensive missiles. Certain technical improvements such as better radar capabilities, employment of optical systems, and an increased number of interceptors would be required in this case and renegotiation of certain provisions of the ABM Treaty would be needed.[13]

Unlike the situation ten years ago, there would likely be no or only limited public protest against BMD deployments on the part of environmentalist groups concerned with the siting of nuclear weapons. Under CSB only 10-15 square miles or so are required to deploy 100 MX capsules, and collocated BMD deployments would occupy limited space. Use of the site at Grand Forks would entail use of current Department of Defense land.

Strategic Employment Doctrine

The doctrinal framework within which the United States negotiated the ABM Treaty was essentially that of the so-called mutual assured destruction school. This school argues that strategic stability in the superpower relationship is best achieved by the capability of each power to hold the other side's population hostage to a retaliatory strike. In this period they argued that 'An ABM system that guards population destabilizes the strategic balance. It reduces the confidence of either side that its own retaliatory force will get through to its targets. The same is true of a system that can be adapted for city defenses.'[14] This statement helps explain why many opponents of missile defense of cities were also opposed to the defense of the ICBM force in this early period. There were a number of critics of the assured destruction persuasion involved in the BMD debate who argued that the planning emphasis should be placed on the assured survival of the U.S. population rather than its assured destruction and that BMD could contribute to that planning objective, but their views were in a clear minority.

In the early 1980s the doctrinal framework is considerably different from that which existed in the earlier period. Two years after the ABM Treaty was signed, the so-called Schlesinger Doctrine was announced as a new departure in U.S. nuclear employment policy, emphasizing greater flexibility in the execution of the Single Integrated Operations Plan (SIOP). The Schlesinger Doctrine provided for a greater planning emphasis on regional and limited nuclear strike options and argued the case for greater counterforce capabilities. It represented a definite trend away from assured destruction thinking toward a greater concern with limiting damage by attacking military as opposed to urban-industrial targets.

Presidential Directive (PD) No. 59, signed by President Carter late in his Administration, was a further elaboration and refinement of the

Schlesinger Doctrine. PD No. 59 stressed the critical importance of targeting Soviet military assets and drew particular attention to the deterrence value of holding the Soviet political and military leadership at risk. Provision was made in the directive for the continued development of limited and selective strike options with a view toward maintaining capabilities for the control of the escalation process. The directive also stressed the significance of enduring command, control, and communications capabilities throughout the period of a possible nuclear exchange and indicated the need to plan for the possibility of a protracted nuclear war. Attention was given in the directive to the requirement for a secure reserve force to protect against possible Soviet attempts to coerce the United States in the post-SIOP period. Fundamental to PD No. 59 was the importance attached to the enduring survivability of strategic forces and the C^3I assets of the National Command Authority.[15]

The Reagan Administration has chosen to endorse the basic provisions of PD No. 59 and current work within the Department of Defense is for the most part focused on implementing the directive. New work in the area of strategic employment policy is directed at identifying the basic characteristics of protracted nuclear wars through scenario construction and developing an over all plan for aligning weapons acquisition policy more directly with targeting doctrine.

Neither the Schlesinger Doctrine nor PD No. 59 dealt directly with BMD. There appears to be an increasing recognition within the community of planners responsible for weapons employment doctrine, however, that BMD considerations could be critical to the further elaboration of current doctrine and could possibly lead to the formulation of new doctrine, one that could possibly be more defense-oriented than offense-oriented.

Arms Control

It may be accurate to say that the defence decision environment of the early 1960s and early 1970s was dominated by high expectations about being able to work out meaningful and lasting arms control agreements with the Soviet Union. SALT I was an unprecedented venture for the two superpowers and many in the United States believed that strategic arms control was the solution to national security.

One critic of BMD argued at the time that it was 'history's judgment' that BMD deployments would adversely affect U.S. national security, that such deployments would accelerate the strategic arms race and jeopardize prospects for arms control agreements.[16] Another critic, later to become the chief SALT negotiator in the Carter Administration, argued at the time the ABM Treaty was under review in the Senate that the Treaty would constitute the single greatest major accomplishment in containing the nuclear arms race; that BMD restrictions would eliminate fears of a side acquiring a first strike capability; and that the Treaty would ensure strategic stability and forestall the accumulation of additional offensive weapon systems.[17]

Many in this early period looked upon the ABM Treaty as an incentive for the Soviet Union to agree eventually to long-term limitations on offensive systems. This viewpoint in fact was implied in Unilateral Statement A made by the U.S. Delegation during SALT I. This statement contains the following two sentences: 'If an agreement providing for more capable strategic arms limitations were not achieved within five years, U.S. supreme interests could be jeopardized. Should that occur, it would constitute a basis for withdrawal from the ABM Treaty.'

In the decision environment of the early 1980s there appears to be a considerably more qualified evaluation of what arms control can in fact contribute to national security. The Reagan Administration has committed itself to a number of arms control initiatives but there is no great expectation in evidence so far that negotiations will lead to an immediate and high pay-off. Some in the Administration are known to favor a policy of extensive arms modernization before serious negotiations get underway. The views of the current Administration have been developed, it should be noted, against the background of many years of unsuccessful SALT II negotiations and lingering allegations that the Soviets have failed to comply in more than one instance with the provisions of the SALT I agreements. It should be evident to the Reagan Administration that, as a further consideration, the ABM Treaty has failed to live up to the claims made by its ardent supporters. The nuclear arms race continues; first strike concerns are greater today than they were a decade ago; and neither strategic stability is ensured nor the build-up of offensive weapons forestalled. Soviet strategic offensive modernization programs have clearly proceeded at a pace that would lead the uninformed observer to believe that the ABM Treaty had never been signed and that Soviet ICBMs and SLBMs faced a formidable BMD deployment in the United States.

Some in the Reagan Administration would be quite prepared to assert in START the linkage between BMD and offensive systems established in SALT I, but no official steps are known to have been taken in this regard. Both the Secretary of Defense and the U.S. Ambassador to the START negotiations have, however, expressed the view that the Reagan Administration is not irrevocably committed to the ABM Treaty.

As has already been noted, the tentative thinking in the U.S. Air Force and Army is that MX deployments in the CSB configuration could be defended at least initially within the constraints of the ABM Treaty.

Therefore, there are no grounds for arguing that near-term BMD deployments would undermine current arms control agreements.

In a related vein, some have argued that prospective BMD deployments could be a potential arms control *initiative*. Their basic point is that layered defense of ICBMs based deceptively in silos might lead to significant strategic force reductions: 'The economics of introducing layered defense using deceptive basing of MX missiles in silos . . . inhibits Soviet expansion of its ICBM force by placing the Soviet forces at a substantial cost disadvantage.'[18] They suggest that the ABM Treaty could at some point be usefully modified or even replaced with other (unspecified) agreements.

Net Assessment and Concluding Observations

One can conclude from the above comparative review that the decision environment of the early 1980s is considerably different from that which existed in the period leading up to the signing of the ABM Treaty. With respect to the definiton of the threat, there is clearly a more discriminating definition today of whose reentry vehicles BMD would be deployed to intercept. Soviet SS-18s and SS-19s and the next generation of those missiles are the weapons of concern to U.S. planners. With respect to deployment policy, there is considerably less ambiguity in today's environment about the scope and level of deployment and whether the basic deployment objective is the defense of ICBMs or some kind of area defense. Deployment plans are basically tailored to the defense of the MX missile deployed in the Closely Spaced Basing configuration or in Minuteman silos.

The decision environment with respect to BMD technology has also changed significantly. However, serious questions continue to be asked regarding this decision area that are not unlike the questions posed a decade ago. The feasibility, reliability, survivability, and susceptibility (to Soviet countermeasures and 'design around' engineering capabilities) of endoatmospheric and particularly exoatmospheric systems remain in many respects uncertain and much work remains to be done before BMD is a sufficiently demonstrated technology. It might be noted here that critics of BMD who tend to rest their case on the argument that complexity is the fatal flaw of BMD systems and/or that operational demands on the systems in a nuclear war would be very different than those imposed under controlled test programs are making basically the kind of critical generalizations common in most debates about modern weapons technologies. The fact is that modern technology *is* complex. A potential enemy could end up paying a terribly high price, however, if that complexity is seen as an incentive rather than as a deterrent to nuclear aggression.

With respect to arms control, the START negotiations underway in Geneva since late June 1982 are dramatic in terms of the kind of reductions being sought by the United States, but the prospects for near-term agreement are not promising. The Soviet position is that the U.S. proposals focus on weapon systems (essentially ICBMs) in which the Soviets have a comparative advantage and ignore systems (bombers and cruise missiles) in which they lag behind the United States. The Reagan Administration has rejected the results of SALT II as the model or primary point of departure for START, but as the negotiations evolve there is likely to be more of a U.S. willingness to look to the SALT II Treaty for ways to move the negotiations forward if that is seen to be in the national interest. The Treaty limits each side to flight-testing and deployment of only one new type of ICBM, and the MX would be one new type flight-tested by the United States. In addition, the Treaty permits deployment of mobile ICBM launchers, and MX deployed in the CSB mode, which constitutes a mobile system, would therefore be consistent with the terms of the Treaty.

Some consideration is being given, however, to the possible advantages of banning mobile ICBM launchers, perhaps the key advantage being the easing of verification requirements. Banning mobile ICBM launchers was first proposed in SALT I and it could appear at some point on the agenda for discussion at START.

Decision making with respect to MX/CSB is complicated by the recognized limitations of other basing mode options for MX: continuous patrol aircraft and deep underground basing.[19] The report of the Townes Panel, appointed in March 1981 to review the MX missile system program and to advise the President and the Secretary of Defense on the best course of action to follow, and chaired by Professor Charles Townes of the University of California, stated that continuous airborne patrol appeared to be the 'most promising approach to providing a new secure ICBM retaliatory force.' The Panel recommended that

> the concept of keeping ICBMs on patrol over oceans and the continental U.S. be pursued to the extent of initiating a program for such an aircraft, and proceeding promptly to concept formulation. Based on initial studies, the fuel-efficiency of such an aircraft appears to be sufficiently high, the technical risk sufficiently low, and the potential survivability (including consideration of countermeasures) sufficiently good to call for a prompt and thorough examination of this basing concept in order to make the earliest possible decision for development and deployment.[20]

Despite this recommendation, very little support for this option developed elsewhere in the policy community, least of all within the Air Force. Technical studies have cited a number of limitations of continuous patrol aircraft: dependence on early warning of attack unless the aircraft were continuously airborne; problems of postattack endurance; and high operational costs.[21]

Deep underground basing was given only limited attention by the Townes Panel, which simply recommended that the concept be promptly and vigorously investigated. A number of studies have in fact been conducted concerning deep underground basing in the past few years and its limitations are relatively well known. Deep underground basing does not meet one of the stated requirements for the MX which is to provide a quick-response capability for time-urgent missions.

In view of the limitations of and general lack of support for both continuous patrol aircraft and deep underground basing, it would appear that CSB will remain the frontrunner among the basing options and may very well be the final choice. A Defense Science Board task force has been formed to conduct an assessment of the CSB concept and is scheduled to report to the Secretary of Defense by September 1, 1982. The Air Force is to make its final recommendation to the Secretary of Defense by the same date. Full scale development of CSB could begin as early as FY 1983.

An adequate discussion of CSB is beyond the scope of this paper, but it can be pointed out that the Air Force has examined various kinds of potential Soviet attack options (full spike, partial spike, walk attack, and pin-down) and remains confident that CSB would be effective through the late 1980s. In the view of the Air Force, the Soviet Union would have to develop new technologies (large-yield reentry vehicles, soft-landers, earth-penetrators, and/or low yield-high accuracy maneuvering reentry vehicles) seriously to threaten MX/CSB.[22]

In making the case for CSB the Air Force does not argue that BMD is required to ensure the survivability of MX in the CSB mode. That argument could call into question the technical integrity of CSB. The Air Force position is that BMD might be required as a follow-on option. At the same time, there are apparently many in the Army who have serious reservations about the advisability of proceeding with CSB and who would prefer to defend MX deployed in Minuteman silos. Defending the Minuteman silos would involve a single-tier defense whereas defense of MX/CSB would involve both endoatmospheric and exoatmospheric interceptors and technologies.

In the event CSB is not selected as the long-term basing mode for MX, the expectation is that plans for basing the missile in Minuteman silos will continue beyond 1982 Fiscal Year and research on both deep underground basing and BMD would also continue. It is not expected that the continuous patrol aircraft option will prove viable even in the near term.

Finally, with respect to strategic employment doctrine, the Schlesinger Doctrine and PD No. 59 represent a definite shift away from the doctrinal environment within which the ABM Treaty was signed ten years ago. BMD has become a potentially significant element in the current doctrinal emphasis on strategic survivability and endurance. It is the case, however, that in the past ten years strategic planners and policymakers alike have largely failed to integrate BMD into long-range strategic war thinking. The linkage between strategic defenses and strategic offenses established in SALT I has been for the most part ignored in the past ten years, and the existence of the ABM Treaty is too often cited as the reason, or excuse, for not coming to grips with defensive possibilities. Problem-solving at the strategic level has been geared almost exclusively in the past decade to finding solutions through offensive modernization programs alone, and offensive force structure planning and targeting analyses typically isolate BMD as some kind of independent force element. Technology should not determine doctrine. Rather is it the responsibility of planners and policymakers to design doctrine, taking into account technological possibilities but not being captive to those possibilities.

What the foregoing suggests is that ballistic missile defense appears to be in greater alignment with the decision environment of the early 1980s than it was a decade ago. However, the story remains incomplete and more time is required to see if the various decision areas will converge to a point where a sufficiently convincing case can be made that the United States should drastically revise its arms control position on ballistic missile defense. There appears to be no compelling argument for proposing modifications to the ABM Treaty at the second five-year review in 1982, particularly in view of the possibility that MX could be deployed in a mode where BMD would not be relevant. If the decision is made to deploy MX in CSB, deployments of Sentry and Spartan missiles could be planned initially within the constraints of the ABM Treaty. This approach would allow for progress to be made in the START negotiations and would buy time for further evaluation of BMD technologies. Further thought could also be given under this approach to the advisability of developing a more balanced strategic doctrine for the United States in the years ahead, a doctrine that would place much greater emphasis on strategic defense capabilities.[23]

NOTES

1. The terms *decision environment* and *decision area* are borrowed from Kenneth J. Arrow, *The Limits of Organization* (New York: W. W. Norton & Company).
2. The comment is that of Jack Ruina, professor of electrical engineering at the Massachusetts Institute of Technology, in 'ABM Revisited: Promise or Peril?', *The Washington Quarterly*, Vol. 4, No. 4, Autumn 1981, p. 12.
3. Spurgeon M. Keeny, Jr. and Wolfgang K. H. Panofsky, 'MAD vs. NUTS: The Mutual Hostage Relationship of the Superpowers,' *Foreign Affairs*, Vol. 60, No. 2, Winter 1981/82.
4. This and the other quotations in this paragraph are taken from Abram Chayes and Jerome B. Wiesner, eds., *ABM: An Evaluation of the Decision to Deploy an Antiballistic Missile System* (New York: Harper and Bros., 1969), pp. 12-24. Missing so far from today's debate is the kind of criticism leveled against Sentinel/Safeguard by such critics as Herbert York who argued that those systems could be compared to the Maginot Line. See his *Race to Oblivion* (New York: Simon and Schuster, 1970). Winston Churchill saw the Maginot Line differently: 'the Maginot Line must be regarded as a wise and prudent measure. . . . It could . . . have served as a trusty shield, freeing a heavy, sharp, offensive French sword.' *The Second World War. The Gathering Storm*. Vol. 1 (Cassell and Co., Ltd., 1948), p. 426.
5. William A. Davis, Jr., 'Ballistic Missile Defense Will Work,' *National Defense*, Vol. LXVI, No.373, December 1981. Dr. Davis is the Deputy Program Manager of the Army's Ballistic Missile Defense Systems Command at Huntsville, Alabama. The quotations and other information found in this paragraph are taken from Dr. Davis' article. In addition, see the same author's 'Current Technical Status of U.S. BMD Programs,' in *U.S. Arms Control Objectives and the Implications for Ballistic Missile Defense*, Proceedings of a Symposium held at the Center for Science and International Affairs, Harvard University, November 1-2, 1979, pp. 29-53. Exoatmospheric systems are designed for midcourse intercepts and exotic systems refer to such technologies as intense laser beams or particle beams employed for boost-phase or early-trajectory intercepts.
6. Davis, 'Ballistic Missile Defense Will Work,' *op. cit.*, p. 42.
7. G. E. Barasch, D. M. Kerr, R. H. Kupperman, R. Pollock and H. A. Smith, *Ballistic Missile Defense: A Potential Arms-Control Initiative*, Los Alamos National Laboratory, Los Alamos, New Mexico, January, 1981, p. 6.
8. *Ibid.*, pp. 7-8.
9. *Ibid.*, p. 8.
10. These were the issues in the celebrated debate between Albert Wohlstetter, on the one hand, and George Rathjens, Ralph Lapp, and others, on the other hand, in which Wohlstetter's analysis was later judged the more objective by the Operations Research Society of America. Cf. Albert Wohlstetter, 'The Case for Strategic Force Defense,' in Johan J. Holst & William Schneider, Jr. (Eds.), *Why ABM? Policy Issues in the Missile Defense Controversy* (New York: Pergamon Press, 1969), pp. 119-42.
11. This is one of the arguments, for example, of John D. Steinbruner, 'Strategic Vulnerability: The Balance between Prudence and Paranoia,' *International Security*, Summer 1976.
12. Henry A. Kissinger, *White House Years* (Boston: Little, Brown and Company, 1979), p. 542.
13. The information in this paragraph is based on *M-X Closely Spaced Basing*, Department of the Air Force, AF/RD-M, June 28, 1982.
14. Chayes and Wiesner, *op. cit.*, p. 51.
15. The development of U.S. strategic employment policy in the 1970s is discussed in Leon Sloss, 'The Evolution of the Countervailing Strategy,' a paper delivered at the National War College, February 17, 1982. Mr. Sloss was the director of the targeting studies group in the Department of Defence whose work led to Presidential Directive No. 59. Additional information is available in *Discussion of Nuclear Weapon Employment Policy*, a compilation of open-source articles and statements concerning

PD-59 published by the Public Affairs office of the Department of Defense, no date.

16.　Bernard T. Field, 'The ABM and Arms Control,' in Chayes and Wiesner, *op. cit.*, pp. 187-92.

17.　This is a paraphrase of the testimony given by Paul Warnke at the Senate hearings.

18.　Barasch, Kerr, Kupperman, Pollock and Smith, *op. cit.*, p. 23.

19.　The various basing options for the MX missile are reviewed and analyzed in *MX Missile Basing*, Office of Technology Assessment, U.S. Congress, Washington, D.C., September 1981. This particular publication did not consider the CSB concept.

20.　*Report of the Committee on M-X Basing*. Executive Summary. News Release of the Office of Assistant Secretary of Defense (Public Affairs), March 23, 1982, p. 2. This document was originally classified. The judgment of the Townes Panel with respect to BMD was relatively categorical but left room for future possibilities: 'There is no demonstrated technology or system of sufficient performance to warrant commitment today to a Ballistic Missile Defense (BMD) deployment to defend ICBMs in silos. However, BMD may play a future role in helping secure our retaliatory forces if some promising concepts materialize, and the Committee recommends expanding the current BMD research and development effort.' *Ibid.*, p. 3. It must be remembered in this connection that CSB emerged as a full concept only after the Panel completed its work.

21.　See, for example, *MX Missile Basing*, *op. cit.*

22.　*M-X Closely Spaced Basing*, *op. cit.*, pp. 4-5 and pp. 10-11.

23.　One must conclude with a historical reminder. In describing some of the National Security Council meetings leading up to the signing of the ABM Treaty, Henry Kissinger said years later that he regretted to admit that at those meetings he was 'swayed by bureaucratic and political considerations more than in any other set of decisions in my period in office.' He also said that at one crucial NSC meeting the discussions 'had all the elusiveness of a Kabuki play. Each department invoked complicated technical arguments in which the same facts were used to produce radically different conclusions. . . . All of this feinting and posturing was performed before a President bored to distraction. His glazed expression showed that he considered most of the arguments esoteric rubbish.' *White House Years*, *op. cit.*, pp. 542-3. In other words, there are no guarantees that the next time around there will be a rational decisionmaking process or competent decisionmakers.

Arms Control Implications of
Ballistic Missile Defense

Robert C. Gray

When the Reagan Administration cancelled plans for the deployment of MX missiles in multiple protective shelters, it announced that a new basing mode would be chosen from among three alternatives: continuous airborne patrol, deep underground basing, and ballistic missile defense.[1] Some analysts have argued that the deficiencies of the first two alternatives virtually guarantee that the third – ballistic missile defense (BMD) – will be chosen.[2] There is, in fact, mounting evidence that the Reagan administration takes BMD[3] quite seriously, despite Secretary of Defense Caspar Weinberger's cautious language about BMD prospects in the *Annual Report* for Fiscal Year 1983.[4] The BMD funding request for FY 1983 has been increased 88% (from $463.6 million in FY 1982 to the proposed $870.5 million)[5]. And high officials of the Department of Defense recently called the Army BMD program 'the hottest game in town.'[6]

The ultimate decision on ABM development and deployment should be based on a net assessment of many factors. Among these are the following: the effectiveness and cost of projected BMD systems as compared with other methods of improving the survivability of ICBMs; an assessment of Soviet offensive and defensive programs; the impact on Western (and Chinese) offensive forces of *Soviet* BMD deployments unconstrained by the ABM Treaty; the implications for deterrence (and U.S. and Soviet strategic doctrines) of BMD deployments with area-defense potential; and the impact on the prospects for future arms control agreements of revising or terminating the ABM Treaty.

The purpose of this study is to focus on only one dimension of such a comprehensive assessment – the arms control implications of BMD development and deployment. Arms control implications, however, cannot be assessed until BMD technologies are first described.

BMD Options

In discussing ballistic missile defense, the question asked so frequently in the ABM debate of 1969-70 must be addressed: will it work?[7] To a great extent, the answer to this question depends upon what mission one is asking BMD to perform. Of the two major missions, defense of hardened targets (such as ICBMs and command and control centers) and defense of areas (such as cities), the former is by far the easier task. The leakage of a single reentry vehicle (RV) through a BMD designed for city defense could result in an unacceptably high level of destruction. As the Deputy Program Manager of the Army BMD effort put it recently, 'ICBM sites are easier to defend than cities because they are hardened against nuclear effects and because they do not have to be defended to a survival level approaching 100 per cent.'[8] For this reason the current American BMD program is largely confined to the mission of defending hardened targets.

One of the most useful ways of characterizing BMD systems is in terms of 'where on the trajectory' of a ballistic missile interception is designed to occur.[9] In this discussion the focus will be on those systems that seem feasible and have some potential for deployment in the 1980s and 1990s. Using these criteria, certain options at either end of the trajectory of a ballistic missile will be ignored. Space-based directed-energy weapons which would be designed to intercept ballistic missiles shortly after launch seem unlikely to be available for deployment in the near-term.[10] At the other end of the trajectory, 'last-ditch' methods which have been proposed for intercepting RVs shortly before impact have failed to attract significant constituencies for development. 'Dust defense,' for example, would involve burying 'clean' nuclear weapons in the vicinity of ICBM silos. Detonated immediately before the arrival of RVs, 'the dust and debris lofted into the air would destroy approaching RVs.'[11] In addition to technical questions about such a system, the likely public opposition to detonation of U.S. nuclear weapons on American soil effectively removes this option from serious consideration.[12] 'Simple' systems that do not involve nuclear weapons have not been attractive to the Army BMD program. As the Deputy Program Manager put it, 'simple BMD systems which meet the criteria of low cost, rapid deployability and adequate effectiveness are difficult to synthesize.'[13]

In describing the BMD systems of greatest policy relevance, a cautionary note is in order. The information currently available is fragmentary and probably biased. As Jonathan Medalia of the Congressional Research Service recently noted, key questions concerning system performance, cost, and time-frame of deployment cannot be fully answered 'because of the security classification of some data, technological complexity, the difficulty of evaluating capability in relation to countermeasures, *and the fact that most recent analysis of BMD has been prepared by BMD advocates.*'[14]

A BMD system must perform four tasks: target-detection, target-recognition, target-tracking, and target-destruction.[15] The Safeguard system used ground-based radars to perform the first three of these tasks as well as to guide the interceptor missiles for target-destruction. The layered Safeguard system employed Perimeter Acquisition Radars (PAR) for target-detection and target-recognition (or early warning and attack characterization). Missile Site Radars (MSR) were used to guide the exoatmospheric Spartan missiles (carrying single nuclear warheads in the several-megaton range) to the point of intercept. RVs leaking through this first layer were to be intercepted by endoatmospheric Sprint missiles, also carrying single nuclear warheads (in the several-kiloton range) and also guided by the Missile Site Radars.[16]

Cautious advocates of current BMD technologies admit the technical deficiencies of the Safeguard system. The nuclear effects of the detonated Spartan warhead would have made it impossible for the MSRs to perform their missions. The radars, large and relatively few in number, were vulnerable to attack. Even if they survived, the radars could not discriminate RVs from decoys in space. Computer technology was not advanced enough to cope with the demands of the system. And the system could have been defeated by any number of Soviet counter-measures.[17]

Proposed BMD systems are being designed to avoid as many of these problems as possible. The most discussed concept is a layered or Overlay system consisting of an exoatmospheric component and an endo-atmospheric Low Altitude Defense System (LoADS).

The exoatmospheric component is in a much earlier phase of development than LoADS, and any assessment is necessarily tentative. Current concepts would attempt to overcome the nuclear effects problem faced by Safeguard by using interceptor missiles armed with multiple non-nuclear kill (NNK) vehicles. The large and vulnerable ground-based radars would be replaced by 'fly along computers.'[18] Long-wave infrared (LWIR) sensors would be used to track the threat after receipt of tactical warning of launch from space-based sensors.

A pro-BMD Los Alamos report admitted that the computer requirements of the above system 'stress the current state of the art,' that 'homing and non-nuclear warhead lethality' are 'two elements of NNK that do stretch the limits of present technology,' and that 'significant upgrading' would be necessary for the Overlay to 'be integrated with an already strained command-communications-control system.' Nonetheless, the Los Alamos analysts are optimistic, for in their view 'the technology base for exoatmospheric intercept appears to be either in hand or on the immediate horizon.[19]

Other analysts are less optimistic. Jack Ruina emphasizes that 'an effective overlay system is only in planning stages . . . Some healthy skepticism is in order before major policy decisions are made on the basis of what is currently known.' Jonathan Medalia reports that some critics believe incoming RVs surrounded by balloons 'could simultaneously deceive both radar and LWIR in space' and that 'a massive Soviet attack' would simply overwhelm the Overlay. And the Office of Technology Assessment concluded that 'for the moment, facing the relative immaturity of the Overlay concept and a near-term decision regarding MX basing, it would be quite risky to rely on Layered Defense [the Overlay] as the basis for ensuring MX survivability.'[20]

The endoatmospheric level of the Overlay would be provided by LoADS or a derivative thereof. Designed to be deceptively based in the multiple protective shelter (MPS) configuration of the MX missile, LoADS is currently being reoriented toward providing low altitude defense of 'ICBMs that are either fixed or deceptively based in any of the deployment schemes being considered.'[21] Although specific characteristics of the system would be determined by the basing mode, the general characteristics of endoatmospheric BMD technology can be assessed by examining LoADS as it was developed for MPS basing.

Unlike the endoatmospheric Sprint component of the Safeguard program, which was designed to operate between altitudes of 100,000 and 300,000 feet, LoADS is designed to operate below 50,000 feet. Also unlike Sprint, which was dependent on a few large, vulnerable ground-based radars and ground-based computers, LoADS has been designed to consist of numerous deceptively-based units, each containing phased-array radars, small computers, and interceptor missiles. As designed for the defense of MX-MPS, the LoADS interceptors would have employed a nuclear warhead of low yield.[22]

The base-line deployment plan for MX in the multiple protective shelter mode was one missile for each cluster of 23 shelters, for a total of 200 missiles and 4600 shelters. Since the system was designed so that the Soviets would not know which shelter contained the missile, each shelter would have to be targeted. Thus, to destroy 200 missiles, the Soviets would have to expend 4600 warheads.

If the number of accurate Soviet RVs increased, the deployment of 200 LoADS defense units might well have been chosen over additional shelter construction, for the coupling of BMD with deceptive basing would have provided great leverage. A LoADS defense unit programmed preferentially to defend the shelter containing the MX missile would have forced the Soviets to assume that the first RV targeted on that shelter would be intercepted. To destroy 200 missiles, the Soviets would have had to double target each shelter, expending 9200 warheads. It is this exchange ratio that made LoADS appear so promising.[23]

The Los Alamos analysts judged LoADS to be 'a relatively low technical risk system' deployable (presumably for MX-MPS) 'by the mid-eighties.' They also asserted that 'a heavy LoADS system [some 1800 interceptors deceptively deployed] could be effective as a stand-alone defense of Minuteman silos starting by 1987.' Referring only to the MX-MPS role, the Office of Technology Assessment concluded that despite such challenges as low-altitude intercept and the ability of LoADS to operate in an extremely stressful nuclear environment, 'there are [no problems] which should stop LoADS from doing its job as well as it needs to.'[24]

Critics of LoADS describe measures that could be adopted by the Soviets to overwhelm the system. These include the deployment of maneuvering reentry vehicles, the possible detection of the shelter containing the missile in the MPS configuration, and an increase in RVs to such a point that LoADS could be saturated. With reference to a LoADS defense of Minuteman, some have suggested that 'it might be difficult to obtain enough nuclear material for the 1800 warheads.'[25]

Because of the uncertainties concerning the basing mode of MX, the Army BMD program is avoiding commitments to any one defensive concept. Efforts are underway to provide a terminal defense system based on LoADS for fixed *or* deceptively based ICBMS. And 'a long-term effort' is underway to provide the exoatmospheric system that would make a layered defense (or Overlay) possible.[26]

Will ballistic missile defense work? Keeping in mind the fact that available sources are less than adequate for a definitive judgement, it *appears* that technical prospects are brighter for the defense of hardened targets than they were for Safeguard a decade ago. Hence, fairly vigorous research and development should be encouraged. Indeed, this would be the prudent course as a hedge against *Soviet* BMD efforts even if there were little U.S. enthusiasm for strategic defense. Care should be taken, however, to ensure that a decision to move to engineering development or, thereafter, to deployment, is based on a careful assessment of all relevant factors and not primarily on the momentum of technology.

The Constraints of Arms Control: The ABM Treaty

The only component of the SALT-I agreement legally binding at this time is the Limitation of Anti-Ballistic Missile Systems. Signed and ratified in 1972, the Treaty was designed to prevent either side from deploying (or providing the basis for deploying) 'ABM systems for a defense of the territory of its country' except as provided for by the Treaty (Article I). The Arms Control and Disarmament Agency has stated that 'each country thus leaves unchallenged the penetration capability of the other's retaliatory missile forces.' The Treaty initially limited the U.S. and the U.S.S.R. to two ABM sites, one to protect the national capital and the other to protect ICBM fields. (A 1974 protocol amended the Treaty to limit each side to only one site.)[27]

Setting quantitative limits, the Treaty specified that each ABM site could contain no more than 100 interceptor missiles and 100 launchers. The number of radars at each site was also limited (Article III). Setting qualitative limits, the Treaty constrained, among other things, the development, testing, or deployment of 'sea-based, air-based, space-based, or mobile land-based' systems as well as systems for multiple launch or rapid reload (Article V). (An Agreed Statement stipulated that Article V prohibited MIRVed warheads as well.) ABM systems based on physical principles other than those constrained in the Treaty were to be 'subject to discussion' between the two countries. Neither side could impede verification by national technical means (Article XII).

A Standing Consultative Commission (SCC) was established to deal with such issues as ambiguous situations developing under the Treaty. The Treaty provided for regular five-year reviews, although either party could propose amendments at any time (Article IV). In addition, either side could, on six-months' notice, withdraw from the Treaty 'if it decides that extraordinary events related to the subject matter of this Treaty have jeopardized its supreme interests' (Article XV).

The Soviet Union maintains the Galosh system to defend the Moscow area, although it is generally recognized that this system is ineffective. The United States deployed the Safeguard system to defend ICBMs near Grand Forks, North Dakota, although with the exception of two radars, this was dismantled following a Congressional vote in 1975.[28] Both countries have maintained research and development programs on ballistic missile defense. A number of factors have converged to raise questions about the continued viability of the ABM Treaty. Prominent among these is the issue of how constraining the Treaty will be on the continued (indeed, accelerated) development of the technologies described in the previous section. The issue has been publicly joined of late because the second five-year review of the Treaty is scheduled for the fall of 1982, and many have wondered whether the U.S. might take this as an opportunity to amend or abrogate the Treaty.

Some analysts viewed President Reagan's cancellation of the MPS basing mode as a setback for near-term deployment of BMD and, hence, as a reprieve for the ABM Treaty. These analysts argued that the only immediate reason to seek changes in the ABM Treaty would have been to prepare the way for LoADS defense of MX in the MPS mode. Deployment of LoADS in defense of silo-based MX or deployment of any other BMD system was deemed to be too distant a prospect to necessitate action on the Treaty in 1982. Hence, one observer remarked recently that 'President Reagan's decision to abandon MPS would appear to relieve the near-term pressure for a revision of the ABM Treaty.'[29]

The principal *technical* issue pertaining to the continuation of the ABM Treaty in the near-term (1982-1987) is whether, within the terms of the Treaty, research, development, and testing can be pursued vigorously enough to support future BMD deployment options. It is clear that no militarily useful *deployments* could occur within the '100 missile/launcher' and geographical constraints of Article III. The most that could realistically be done is deployment of allowed ABM components in a technology validation program. Because crucial elements of the systems under consideration go beyond what is allowed by the Treaty, however, such a deployment probably makes little sense.

The layered defense system (exoatmospheric plus LoADS) would be constrained in a number of ways by the Treaty. Article V would prohibit the 'development, testing, or deployment' of the multiple NNK vehicles as well as any mobile basing of launchers and radars. It is possible that this article also prohibits the 'exoatmospheric [LWIR] probes and air-borne optical platform' envisaged for the Overlay.[30] Article XII may prohibit deceptive basing of BMD as interfering with 'national technical means of verification.'

One of the most restrictive constraints on BMD efforts is the Article V prohibition on *development* (as well as testing and deployment) of certain types of systems. 'Development' of ABMs has been defined as follows:

> The obligation not to develop such systems, devices, or warheads ['sea-based, air-based, space-based, or mobile land-based' ones as well as MIRVed warheads or 'rapid reload of ABM launchers'] would be applicable only to that stage of development which follows laboratory development and testing. *The prohibitions on development contained in the ABM Treaty would start at that part of the development process where field testing is initiated on either a prototype or bread-board model.*[31]

As the Office of Technology Assessment concluded, use of this definition prohibits development beyond the laboratory of LoADS' 'mobile components' or the exoatmospheric multiple kill vehicles.[32]

Despite such limitations, the official view has been that the U.S. BMD program can, at least in the near-term, be conducted within the framework of the Treaty. In early 1981, before BMD funding was almost doubled by the Reagan administration, James Wade, then Acting Principal Deputy Under Secretary of Defense for Research and Engineering, stated that

With regard to our BMD programs, we can resolve most of the key issues associated with establishing the feasibility of the concepts under consideration within the limits of the Treaty. If we determine that we want to proceed with engineering development of either the LoAD or Overlay concepts it would be necessary to, at some point, negotiate modifications to the Treaty or to withdraw.[33]

Administration statements in support of the greatly increased BMD program submitted in early 1982 have asserted that it is 'consistent with the ABM Treaty.'[34]

The available evidence on R&D requirements in the near-term suggests that there is no pressing need to amend or terminate the ABM Treaty in 1982.[35] Indeed, as William Hyland has observed, 'the notion that 1982 is a major watershed for the ABM is wrong.' Under Article IV of the Treaty, amendments may be proposed at any time at meetings of the Standing Consultative Commission. Hyland has suggested that it would be preferable for any alterations in the Treaty to be first discussed at SCC meetings rather than at the five-year reviews.[36]

The principal arguments in favor of action in 1982 are more likely to be related to political considerations than pressing technical issues. Two analysts associated with the Los Alamos study noted earlier suggested in the fall of 1981 that Treaty review 'needs to be addressed now *even though accelerated technological development would not be constrained by treaty provisions for a number of years* . . . Congress would be unlikely to commit much greater sums to ballistic missile defense with the constraints of the treaty still in place.'[37] This is a particularly insubstantial reason to press for Treaty review in 1982. Such a problem should be addressed when (and if) it occurs. In fact, it seems likely that Congress will follow the responsible course of adequately funding BMD within the confines of the Treaty until such time as amendment or termination is mandated by pressing concerns.

The constraints of the ABM Treaty will be significant if the U.S. decides to press forward in the mid- to late-1980s with systems such as those described in the previous section. These constraints, however, are not serious enough in the near-term to necessitate action to alter or terminate the ABM Treaty in 1982.

Arms Control and Ballistic Missile Defense: Wider Implications

The discussion thus far of BMD technologies and arms control constraints on those technologies is but a small part of a much larger and more complicated picture. If BMD systems become feasible and a U.S. decision is made to amend or abrogate the ABM Treaty, the concept of deterrence on which American strategic nuclear policy has been based for over thirty years may be radically transformed.

The conventional wisdom of most of the arms control and defense community has been the offense-dominant policy of assured destruction.[38] By leaving their countries open to attack by the offensive ballistic missiles of the other, American and Soviet leaders seemed to have codified this concept of deterrence in the ABM Treaty.[39] Many in the arms control community are dismayed by the prospective collapse of this treaty. Recently, the Committee for National Security published *An Arms Control Agenda for the Eighties*.[40] Because it succinctly states the standard arms control case against BMD, its principal points are worth examining:

(1) The Soviets will not renegotiate the Treaty to allow hard point defense to solve the U.S. ICBM vulnerability problem, so the only recourse the U.S. is likely to have is abrogation.

(2) Even if the Soviets negotiated with the U.S., 'it would be impossible to define limitations that would assure the United States that the Soviets were not deploying an area defense system to protect its cities.'

(3) The ABM Treaty is 'the most important arms control agreement in effect today' and changes in or termination of it 'would make difficult, if not impossible, the negotiation of any strategic arms control arrangements . . . in the foreseeable future.'[41]

The first assertion – that the Soviets will not renegotiate so as to allow the U.S. to deploy a defense of its ICBMs – is an oversimplification, for the Soviets might be willing to renegotiate for a price. The key point is that the U.S. would be unlikely to pay the price the Soviets would seek to exact. The Soviet concept of deterrence emphasizes damage limitation. Thus, it is not implausible to believe that Soviet deployment of ICBMs with RVs of sufficient yield and accuracy to threaten U.S. ICBMs was intended to implement a strategy based on this concept.[42] One need not believe, however, that the Soviets are pursuing a 'warwinning' strategy in order to conclude that the Soviets are unlikely to help the U.S. solve its land-based missile problem.

It is not clear that the only outcome of a U.S. proposal for hard point defense would be treaty abrogation, for the Soviets might welcome area defense BMD (as well as hard point BMD) as a contribution to a damage limitation policy. It is presumably in anticipation of such a situation that the Committee for National Security provided a second argument – that even if the Soviets negotiated, it would be impossible to limit each side to unambiguously hard point layered defense.

This assertion is probably correct. The Soviet Galosh system is a layered one, and follow-on systems would probably be similar.[43] If the U.S. sought an amendment providing for layered defense of ICBMs, it would be highly unlikely that the Soviets would forgo the right to deploy a layered system. The crucial problem is that layered systems, unlike LoADS, can be upgraded from hard point to area defense, and both sides would be concerned about this.[44]

An attempt to deploy a layered defense unambiguously designed for hard point defense is probably doomed to failure. From the viewpoint of mutual assured destruction, the price of this failure would be high. As Herbert Scoville put it,

> an ABM defense of ICBM launchers must not simultaneously provide a nationwide defense because this could decrease confidence in both the ICBM and SLBM legs of the deterrent triad, and there would be a net loss of crisis stability . . . [Yet] the exoatmospheric [or layered] system of necessity would almost certainly provide wide area coverage and give rise to fears that the system could be expanded to provide nationwide coverage.[45]

The third argument – that changes in or termination of the ABM Treaty would doom subsequent arms control negotiations – is more problematical than the first two. A number of analysts have argued that BMD development and/or deployment would *enhance* arms control. Jan Lodal, for example, has argued that a serious BMD program designed to provide 'limited defense' would 'provide the best possible "bargaining chip".'[46] The Los Alamos report cited earlier concluded that 'a cost-effective defense – one capable of countering growth in the offense – would intrinsically support arms limitation by eliminating incentives for offense proliferation.'[47]

There are widely divergent views of the impact of BMD on achieving deep cuts in the offensive arsenals of both sides. Lodal has argued that ABM deployments would make it more difficult to negotiate 'lower limits on the sizes of both sides' missile forces.' His position is that each country would want to preserve the 'assured destruction capability' of penetrating the ABM systems of the other. This would require 'relatively large MIRVed missile forces' and 'would make it unlikely that either side would be interested in negotiating reductions to low levels.'[48] Graybeal and Goure provide support for this general view, albeit from a somewhat different perspective.[49]

Keith Payne, on the other hand, focuses on the problem posed by reductions to such low levels that each side fears that a marginal change in the balance could put the offensive arsenal at risk. Payne recommends 'the full exploitation of both active and passive defense,' which he views as consistent with 'significant reductions in offensive force levels.' Indeed, he argues that emphasis on active and passive strategic defense may 'be necessary to reduce the heightened sensitivity to covert noncompliance that would be associated with genuine offensive force-level reductions.'[50]

Even allowing for differences in terms of reference and presuppositions, the divergent positions outlined above illustrate the intellectual ferment (some would call it chaos) that is likely to accompany a renewed debate over ABM. The issue of BMD deployment and the ABM Treaty involves much more than a decision to deploy new weapons systems. As indicated earlier, it involves nothing less than a wholesale alteration in the framework of deterrence. It would be surprising indeed if a shift from an offense-dominant to a defense-dominant world were contemplated, let alone actively pursued, in the absence of intellectual vertigo.

It is difficult to judge with any precision how difficult renegotiation of the ABM Treaty would be, for much would depend on the context of the effort, on Soviet assessments of how BMD would affect their strategic prospects, and on other situational variables. Some tentative observations can nonetheless be made.

The U.S. would most likely face a serious problem in seeking amendments to the Treaty. Unless one postulates a radical shift in U.S. policy toward large-scale active and passive defenses (a postulate militated against by substantial technological and operational obstacles), U.S. goals are almost certain to be at odds with Soviet interests. If, for example, the U.S. sought an amendment securing the right to deploy unambiguously hard point layered defense of ICBMs, the price, as noted above, might well be the deployment in the Soviet Union of layered systems with area-defense potential. Faced with this prospect, the U.S. might indeed decide that no useful purpose was served by maintaining the Treaty. Such a decision would be especially likely if little progress had been made on offensive force limitations.[51]

A strong argument can be made that arms control efforts would not survive the collapse of the ABM Treaty. Although it is *possible* that the elimination of restrictions on strategic defenses would lead the two sides to negotiate reductions of offensive weapons, the argument for this proposition is rather tenuous. The prospect of an offensive *and* defensive arms race constrained only by the resources and will of the two sides seems at least equally plausible. The implications of this scenario for crisis stability, the Atlantic Alliance, and American domestic politics should be pondered as one considers elimination of restraints on ABMs.

It may be, as some have argued, that little progress in arms control can be expected in the next decade no matter what happens to the ABM

Treaty. Such a prognosis, based as it necessarily must be on subjective judgments about the future, should not be used to justify abandonment of the effort to enhance security and preserve peace by way of verifiable arms control agreements.

Conclusion

The intentions of the Reagan administration with regard to BMD and the ABM Treaty are unclear. Secretary Weinberger has expressed uncertainty about the prospects for BMD. Yet funding for the program has been substantially increased, and it appears to be a strong candidate to play a role in MX missile basing.

The accelerated Army BMD program seems to be consistent with the ABM Treaty, but there are subtle signs of reduced commitment to the Treaty. In the first set of arms control impact statements submitted to Congress by the Reagan administration, the language in a number of passages on BMD has been changed from that of the last statements submitted by the Carter administration.[52] The following is but one example:

> Carter statements: The United States fully supports the ABM Treaty (FY 1982, p. 193).
> Reagan statements: The United States continues to be a party to the ABM Treaty (FY 1983, p. 139).[53]

While such a change by itself might be trivial, idiosyncratic, or even inadvertent, it takes on slightly added importance when viewed in light of *Aviation Week and Space Technology*'s recent statement that 'there is a consensus within the arms control community and the Defense Department that the antiballistic missile defense treaty with the Soviet Union is unlikely to survive in its present form.'[54]

If the Treaty is revised, specifics will matter. The nature of the new ABM debate will depend in part on *when* changes in the Treaty are proposed, *how* they are proposed, and *why* they are proposed. If changes were proposed in 1982 amidst a widespread impression that the time was premature, the debate would most likely be more acrimonious (and less constructive) than if changes were proposed later. If changes were proposed with maximum fanfare at a Treaty review, it might be more difficult to negotiate with the Soviets than if changes were proposed quietly at a non-Treaty review meeting of the Standing Consultative Commission. If changes were proposed in order to dismantle a vestige of detente rather than to remove legal prohibitions on the development of systems deemed vital to national security, those proposals would be unworthy of support.

It is possible to imagine a defense-dominant world in which active and passive strategic defenses and greatly reduced offensive arsenals provide greater stability and security than the offense-dominant world we now inhabit. But as Michael Nacht and Herbert Scoville have concluded, the problem of transition to that world poses great, if not insuperable obstacles.[55]

On balance, Albert Carnesale's judgment remains sound:

> The ABM Treaty, with its severe constraints on ballistic missile defense, represents a joint Soviet-American selection of a world in which neither side has a meaningful defense. To move from this non-BMD world to one in which both sides have extensive BMD deployments would be to gamble for high stakes: one possible outcome could be enhanced strategic stability, but another could be disastrous instability. The burden of proof rests on those who would have us make the move.[56]

While those on whom the burden of proof rests construct their case, research and development on BMD should move forward, as should a continually updated net assessment of the factors enumerated at the beginning of this study. Arms control implications are one important set of factors. But the decision on whether or not to deploy ballistic missile defense should rest neither on theological commitments to arms control nor on blind surrender to technological momentum. Rather, like all such decisions, it should rest on a sober determination of what national strategy, weapons deployments, and employment plans are most likely to preserve deterrence.

NOTES

1. For the Reagan announcement, see *New York Times*, 3 October 1981. The full text of the statement distributed to the press is reprinted on p. 12.
2. For one such view, see Donald M. Snow, 'MX Smokescreen,' *Bulletin of the Atomic Scientists*, March 1982, p. 45. For comparative analyses of air mobile, deep underground basing, and ballistic missile defense, see, respectively, pp. 217-32, 269-74, and 111-43 in U.S. Congress, Office of Technology Assessment, *MX Missile Basing* (Washington: U.S., Government Printing Office, 1981).
3. In this article, the terms ballistic missile defense (BMD) and antiballistic missile (ABM) will be used interchangeably except when referring to the 1972 treaty. Some analysts use BMD to refer to the current military program and ABM to refer only to the treaty.
4. In that report, Weinberger stated that 'although ground-based deployment of MX ultimately may require a BMD for survivability, today's BMD technology is not adequate to defend against Soviet missiles. For the future, we are not yet sure how well ballistic missile defenses will work; what they will cost; whether they would require changes to the ABM Treaty; and how additional Soviet ballistic missile defenses . . . would affect U.S. and allied offensive capabilities.' *Annual Report to the Congress: Fiscal Year 1983*, 8 February 1982, p. III-65.

5. For the FY 1982 appropriation, see Jonathan Medalia, 'Antiballistic Missiles,' Library of Congress, Congressional Research Service, 11 January 1982, p. 19. For the FY 1983 proposal, see 'Statement by Major General Grayson D. Tate, Jr.,' Subcommittee on Defense Appropriations, Committee on Appropriations, U.S. Senate, 31 March 1982, p. 5 (mimeographed). General Tate, BMD Program Manager, asserted that the increase for FY 1982 is necessary 'to support the decision in July 1983 regarding U.S. strategic deployment options and the potential IOC for BMD . . . [and to support] the requirement to develop a flexible BMD system design capable of defending ICBMs in either the fixed or deceptive-basing modes under consideration.'

6. Clarence A. Robinson, Jr., 'Emphasis Grows on Nuclear Defense,' *Aviation Week and Space Technology*, 8 March 1982, p. 27.

7. For two interesting volumes setting forth the case for and against ABM deployment see, respectively, Johan J. Holst and William Schneider, eds., *Why ABM? Policy Issues in the Missile Defense Controversy* (New York: Pergamon, 1969) and Abram Chayes and Jerome B. Wiesner, eds., *ABM: An Evaluation of the Decision to Deploy an Antiballistic Missile System* (New York: Signet: 1969).

8. William A. Davis, Jr., 'Ballistic Missile Defense Will Work,' *National Defense*, December 1981, p. 16.

9. This discussion is based on the Los Alamos National Laboratory, 'Quick Look Assessment of Ballistic Missile Defense,' reprinted in *Congressional Record*, 6 June 1980, pp. S 6429-S 6436. See especially p. S 6431. Cited henceforth as 'Quick Look Assessment.' Loren Thompson has argued that this favorable assessment of BMD prospects probably 'influenced' Reagan administration officials who sought increased BMD funding in 1981. 'Backgrounder on ABM and Ballistic Missile Defense,' *WG Backgrounder Series*, The House Wednesday Group, May 20, 1981, p.21 (mimeographed).

10. Directed-energy weapons could be of two major types, lasers and particle-beams. Whether the latter are feasible remains questionable. Thus, because there is greater optimism about the weapons potential of lasers, and because laser weapons may play an anti-satellite role, the American R & D effort has tilted in that direction. For two enthusiastic reports on the potential of laser weapons, see Senator Malcolm Wallop, 'Opportunities and Imperatives of Ballistic Missile Defense,' *Strategic Review*, VII (Fall 1979), pp. 13-21 and Clarence A. Robinson, Jr., 'Laser Technology Demonstration Proposed,' *Aviation Week and Space Technology*, 16 February 1981, pp. 16-19. For more skeptical assessments of laser weapons, see 'A Cooler Look at Laser Weapons,' *Science*, 9 January 1981, pp. 148-49 and Kosta Tsipis, 'Laser Weapons,' *Scientific American*, 245 (December 1981), pp. 51-57. For a critique of particle-beam weapons, see John Parmentola and Kosta Tsipis, 'Particle-Beam Weapons,' *Scientific American*, 240 (April 1979), pp. 54-65.

11. *MX Missile Basing*, p. 126.

12. For assessments of this and other 'last-ditch' systems, see ibid., pp. 126-28 and Richard L. Garwin, 'Effective Military Technology for the 1980s,' *International Security* 1 (Fall 1976), pp. 50-77. (Garwin also discusses lasers.)

13. William A. Davis, Jr., 'Current Technological Status of U.S. BMD Programs,' In *U.S. Arms Control Objectives and the Implications for Ballistic Missile Defense*. Proceedings of a Symposium held at the Center for Science and International Affairs, Harvard University, 1-2 November 1979, p. 43. Henceforth cited as *U.S. Arms Control Objectives*.

14. Medalia, 'Antiballistic Missiles,' p. 6. Emphasis added.

15. SIPRI, *World Armaments and Disarmament: 1981* (London: Taylor and Francis, 1981), p. 267. Different authors describe these tasks in different terms. See, for comparison, the six functions enumerated by Davis in 'Ballistic Missile Defense Will Work,' p. 21 and the six requirements listed in the Los Alamos 'Quick Look Assessment,' p. S 6431.

16. This description of the Safeguard system is drawn from the Los Alamos, 'Quick Look Assessment,' ibid., p. S 6431.

17. Ibid. For an interesting discussion of the impact of Safeguard on American awareness of the problem of electromagnetic pulse, see William J. Broad, 'Nuclear Pulse: Awakening to the Chaos Factor,' *Science* 213, 29 May 1981, pp. 1009-12. The EMP problem is discussed further in subsequent issues of *Science*: 5 June 1981 (pp. 1116-20) and 12 June 1981 (pp. 1248-51).

18. It would still be necessary to have 'ground-based battle-management computers for attack assessment and commitment of interceptors.' These computers and associated command, control, and communications would have to be 'immune to disruption by precursor SLBM attack.' 'Quick Look Assessment,' p. S 6431. *MX Missile Basing*, p. 132. These two documents contain succinct descriptions of the current concepts of systems designed for the exoatmospheric role.

19. 'Quick Look Assessment,' pp. S 6432-33. In congressional testimony in early 1981, Dr. Seymour Zeiberg, the Deputy Under Secretary of Defense for Research and Engineering (Strategic and Space Systems), described recent efforts bearing on the exoatmospheric mission. 'We have enough experiments to know that we can get there, we can detect, we can track, and we can home. *The fundamental question here is whether we can do this in a war environment.* We have some very major experiments coming up [deleted] where we will shoot some of these interceptors against reentry vehicles . . . Following that we will have to prove that this can be done on a very large scale to assure that we can handle large numbers of targets.' Emphasis added. U.S., Congress, Senate, Committee on Armed Services, *Hearings on S. 815, Department of Defense Authorization for Appropriations for Fiscal Year 1982, Part 7*, 97th Cong., 1st Sess., 1981, pp. 4121-22. Zeiberg was presumably referring to the Homing Overlay Experiment (HOE).

20. Jack Ruina, 'BMD Technology,' in the symposium 'ABM Revisited: Promise or Peril?', *The Washington Quarterly* 4 (Autumn 1981), p. 64; Medalia, 'Antiballistic Missiles,' p. 8; *MX Missile Basing*, p. 132.

21. 'Statement by Major General Grayson D. Tate,' 31 March 1982, p. 2.

22. 'Quick Look Assessment,' p. S 6434 and *MX Missile Basing*, pp. 114-18. It should be noted that for defense of MX in the MPS basing mode, LoADS was deemed sufficient. The Overlay becomes necessary for other basing modes.

23. *MX Missile Basing*, p. 119.

24. 'Quick Look Assessment,' p. S 6434. *MX Missile Basing*, p. 121. For a more detailed discussion of the challenges involved in hardening the LoADS DU, see pp. 122-24.

25. These criticisms are reported in Medalia, 'Antiballistic Missiles,' p. 7.

26. 'Statement by Major General Grayson D. Tate,' 31 March 1982, pp. 2-4. General Tate also announced 'a major effort' to develop *endo*atmospheric nonnuclear kill (p. 4). *Aviation Week and Space Technology* described the overall plan as follows: 'The [LoADS] interceptor missile would be armed initially with nuclear warheads in the 5-kiloton yield range for endoatmospheric kills. The system would have an initial operational capability of 1988 to defend MX ICBMs. This interim ballistic missile defense system would be followed by an interceptor missile armed with a non-nuclear warhead for endoatmospheric use, and an exoatmospheric non-nuclear interceptor would follow in the late 1990s' ('Emphasis Grows on Nuclear Defense,' 8 March 1982, p. 28). For arguments by individuals knowledgeable about BMD that endoatmospheric nonnuclear kill is far from being at hand, see Dr. Seymour Zeiberg in *Hearings on S. 815*, p.4121 (complete citation at note 19), and Senator Pete V. Domenici, 'Toward A Decision on Ballistic Missile Defense,' *Strategic Review* X (Winter 1982), p. 25.

27. For the text of the 1972 Treaty and associated statements and understandings, see 'Treaty Between the United States of America and the Union of Soviet Socialist Republics on the Limitation of Anti-Ballistic Missile Systems,' U.S. Arms Control and Disarmament Agency, *Arms Control and Disarmament Agreements: Texts and Histories of Negotiations: 1980 Edition* (Washington: U.S. Arms Control and Disarmament Agency, 1980), pp. 139-47. For the 1974 Protocol, see pp. 162-63. The Protocol entered into force in 1976. The ACDA description of the Treaty appears on pp. 137-38.

28. Mark M. Lowenthal, 'The 1982 Review of the ABM Treaty,' Library of Congress, Congressional Research Service, 3 March 1982, p. 5. The number of Soviet launchers in the Galosh system was recently reduced from 64 to 32.

29. Ibid., p. 11. Also see Jonathan E. Medalia, 'The Reagan Plan for U.S. Strategic Forces: Issues for Congress,' Library of Congress, Congressional Research Service, 4 November 1981, pp. 7-8.

30. E. C. Aldridge, Jr. and Robert L. Maust, Jr., 'SALT Implications of BMD Options,' in *U.S. Arms Control Objectives*, p. 62.

31. Statement by Ambassador Gerard Smith to the Senate Armed Services Committee, as quoted in U.S., Congress, House and Senate, Committees on Foreign Affairs and Foreign Relations, *Fiscal Year 1982 Arms Control Impact Statements*, 97th Congress, 1st Sess., 1981, p. 195. Also quoted in *MX Missile Basing*, p. 142. Emphasis added.

32. *MX Missile Basing*, p. 142.

33. *Hearings on S. 815*, p. 4180.

34. U.S., Congress, House and Senate, Committees on Foreign Affairs and Foreign Relations, *Fiscal Year 1983 Arms Control Impact Statements*, 97th Cong., 2nd sess., 1982, p. 145. Also see 'Statement by Major General Grayson D. Tate,' p. 4.

35. Michael Nacht has come to the same conclusion: 'Even if the [LoADS] system proves to be a very attractive option, it would be premature for the United States to initiate renegotiation of the ABM treaty at the 1982 treaty review meeting of the SCC.' 'ABM ABCs,' *Foreign Policy* 46 (Spring 1982), p. 171.

36. 'If the United States is interested in ABM revision, then the best way to start it would be through some quiet discussions at the . . . SCC.' 'The Soviet View,' in the symposium, 'ABM Revisited,' p. 69. Raymond Garthoff concurs. See ibid., p. 72.

37. Donald M. Kerr and Guy Barasch, 'BMD Technology,' in the symposium 'ABM Revisited,' p. 62. Emphasis added. Others would like to terminate the Treaty because of a belief that 'the most effective way to signal a tough stance toward the Soviet Union would be to dismantle the entire structure of negotiated arms control.' Nacht, 'ABM ABCs,' p. 155.

38. With the changes in declaratory and force employment policy in the late 1970s (culminating in the formulation of the countervailing strategy and the issuance of Presidential Directive 59), U.S. policy has come to rely on much more than assured destruction. However, elements of assured destruction remain as one component of the deterrent posture. For a description of U.S. policy under President Carter, see Harold Brown, *Annual Report to the Congress: Fiscal Year 1982*, 19 January 1981, pp. 37-45. There is no comparable statement of policy under President Reagan, although there is every reason to assume substantial continuity on the premises of U.S. strategic nuclear policy as outlined by Secretary Brown.

39. For the view that the Soviets operate on the basis of a quite different concept of deterrence, see Colin S. Gray, 'A New Debate on Ballistic Missile Defence,' *Survival* XXIII (March/April 1981), pp. 60-71. Also see Jacquelyn K. Davis et al., *The Soviet Union and Ballistic Missile Defense* (Cambridge, MA: Institute for Foreign Policy Analysis, 1980). For the contrary view that the Soviets share important Western concepts, see Garthoff, 'The Soviet View,' in the symposium, 'ABM Revisited,' pp. 66-72.

40. The Committee for National Security, *An Arms Control Agenda for the Eighties* (Washington, 1981).

41. Ibid., pp. 5-7.

42. Sidney Graybeal and Daniel Goure, 'Soviet Ballistic Missile Defense (BMD) Objectives: Past, Present, and Future,' in *U.S. Arms Control Objectives*, p. 87.

43. Aldridge and Maust, 'SALT Implications,' in ibid, pp. 61-62.

44. On the 'intrinsic capability of area defense' of the 'exoatmospheric component of a layered defense,' see Thompson, 'Backgrounder on ABM and Ballistic Missile Defense,' p. 20.

45. Herbert Scoville, Jr., 'The Arms Control Implications of New Ballistic Missile Defense Technologies,' in *U.S. Arms Control Objectives*, pp. 105-6.

46. Lodal makes this recommendation not because he views BMD *solely* as a bargaining

chip but because he believes that if negotiations fail, 'the program is the best response we are likely to have.' 'Deterrence and Nuclear Strategy,' *Daedalus* 109 (Fall 1980), p. 174. For a critical examination of the bargaining chip approach to arms control, see Robert J. Bresler and Robert C. Gray, 'The Bargaining Chip and SALT,' *Political Science Quarterly* 92 (Spring 1977), pp. 65-88.

47.   'Quick Look Assessment,' p. S 6435. Also see G. E. Barasch et al., *Ballistic Missile Defense: A Potential Arms-Control Initiative*, LA-8632 (Los Alamos National Laboratory, 1981). The authors conclude that adoption of a layered or LoADS-only defense 'would create incentives for the Soviet Union to restrain strategic-arms expansion. Mutual arms-control initiatives could follow' (p. 1).

48.   Lodal, 'Deterrence and Nuclear Strategy,' p. 170. There is less incompatibility than initially apparent between Lodal's positive view of ABMs as bargaining chips and his negative view of their impact on prospects for deep cuts. He differentiates between short-run and long-run possibilities for arms control. In the short run he expects 'no more than modest strategic arms control measures throughout the 1980s.' In the longer run, by deploying strategic defenses rather than more offensive weapons, Lodal believes the U.S. 'would find itself ultimately in a stronger position to negotiate effective limits on nuclear deployment' (pp. 170-71). In the longer term, Lodal seems to view significant force reductions as possible.

49.   Graybeal and Goure consider the problem the U.S. would face if the Soviets deployed substantial BMD: 'If we assume continued or even more stringent limitations and reductions of offensive forces – launchers, missiles, and warheads – then the expansion of Soviet defenses could, over time, further increase U.S. difficulties in maintaining an effective assured destruction capability.' 'Soviet Ballistic Missile Defense,' in *U.S. Arms Control Objectives*, p. 88. Almost certainly, recognition of this prospect would impede future limitations on offensive forces.

50.   Keith B. Payne, 'Deterrence, Arms Control, and U.S. Strategic Doctrine,' *Orbis* 25 (Fall 1981), p. 769.

51.   In 1972, the U.S. ambassador to SALT-I, Gerard Smith, made a unilateral statement that 'if an agreement providing for more complete strategic offensive arms limitations were not achieved within five years, U.S. supreme interests could be jeopardized. Should that occur, it would constitute a basis for withdrawal from the ABM Treaty.' *Arms Control and Disarmament Agreements: 1980 Edition*, p. 146. In his confirmation hearings, the current Director of ACDA, Eugene Rostow, emphasized the importance of this unilateral statement. In speaking of the 1982 Treaty review, Rostow said: 'I want to examine what lies behind Ambassador Smith's statement and reconsider it in the light of changing circumstances. But it is obvious that the premises in terms of which the ABM Treaty has been established have not been maintained and no longer exist.' U.S., Congress, Senate, Committee of Foreign Relations, 'Nomination of Eugene V. Rostow,' 97th Cong., 1st sess., 1981, p. 57.

52.   Although arms control impact statements have not met the congressional goal of affecting weapons decisions, they do serve as one indicator of administration attitudes towards arms control. For a case-study that assesses the utility of these statements, see Robert C. Gray, 'The Coordination of Arms Control Policy and the Weapons Acquisition Process: The Case of Arms Control Impact Statements,' *Arms Control: The Journal of Arms Control and Disarmament*, 2 (September 1981), pp. 218-36.

53.   For citations of the FY 1982 and FY 1983 statements, see notes 31 and 34.

54.   Robinson, 'Emphasis Grows on Nuclear Defense,' 8 March 1982, p. 27.

55.   See Scoville, 'The Arms Control Implications of New Ballistic Missile Defense Technologies,' in *U.S. Arms Control Objectives*, p. 110 and Nacht, 'ABM ABCs,' p. 174. As Scoville put it, 'if one nation had an effective system [or thought it did] sooner than the other, an extremely unstable situation would be created, since the former could then threaten the latter without fear of retaliation. The danger of blackmail or even a first strike would be greatly magnified.'

56.   Albert Carnesale, 'Reviving the ABM Debate,' *Arms Control Today*, 11 (April 1981), p. 8.

The ABM and American Domestic Politics

Phil Williams and Stephen Kirby

Introduction

If the mobile basing mode for the MX is out, the Anti-Ballistic Missile may well be in. President Reagan's decision to reject the 'shell game' basing variant for the MX favoured by his predecessor is, on pragmatic if not strategic grounds, a laudable one. It rules out a construction project, the sheer scale of which would have made the pyramids look like a Lego building. Indeed, many assessments of President Carter's plan for the deployment of MX were as concerned about the feasibility of actually constructing it as they were about its survivability when completed. One of the key issues was whether there would be sufficient building material available to meet the requirements of the project without diverting resources from other construction work. Doubts were also expressed – particularly after it became clear that SALT II would not be ratified – that even if the deployment went ahead as planned by the Carter Admin-istration, the mobile MX would still be vulnerable to a Soviet counter-force attack unless protected by ABM. It is clear, therefore, that 'as part of the process of deploying and protecting the MX, ABM has obviously returned as a strategic subject.'[1] This has been further underlined by the rejection of the mobile basing mode, a decision which many observers sympathetic to Ballistic Missile Defence (BMD) regard as offering a major opportunity to press their case. And the case for BMD deployment has already been presented with vigour, skill and enthusiasm. Yet, this should not be surprising: one of the most striking qualities of the Anti-Ballistic Missile is its resilience and durability – politically not strategi-cally. The ABM may have returned so far as the public debate is con-cerned; in terms of development it has never really been away. The Treaty of 1972 was more concerned with limiting deployment than devel-opment, and allowed both superpowers to continue with intensive R & D programmes. Thus it is possible that at some time in the future the 1972 Treaty may appear less as a permanent prohibition on the large scale deployment of an ABM than as a temporary obstacle to the internal pressures for such a move.

Indeed, there may be something almost inevitable about the deploy-

46

ment of an extensive BMD system, given the vested interests involved. This is not to ignore the fact that in the past the ABM has generated considerable opposition. In the late 1960s and early 1970s in particular the issue became highly controversial and there was widespread opposition, especially in the Senate, to the proposal to deploy a system which, the critics argued, was unnecessary, unworkable, and unpopular. To some extent this argument erupted as a result of a particular combination of circumstances which may be unique to the period. Nevertheless, there are some indications that another round of controversy could be precipitated by a major decision to upgrade the programme. Although there is nothing inherently controversial about BMD, the circumstances of the 1980s may not be as propitious for BMD deployment as the proponents of such a move may have hoped. In the long term, however, the demands of the ABM supporters may well prove irresistible. Although it was the critics of ABM who were most vociferous during the four years preceding SALT I, the persistence of the system's proponents should not be underrated. The enthusiastic support which ABM has consistently received from sections of the military, from industry and from certain groups in Congress is particularly striking. Indeed, the constituency favouring ABM was so strong in the mid-sixties that Secretary of Defense McNamara – in a famous (and much analysed) statement in San Francisco on September 18, 1967 – was compelled to announce deployment of ABM, albeit a thin system, despite his very considerable doubts about either its efficacy or its wisdom. This paper attempts to delineate the pressures which brought ABM to the point of deployment at that time. It then examines the ensuing controversy, suggesting that what appeared as a dispute over the Anti-Ballistic Missile was, in large part, a dispute over other things: the ABM provided the occasion for a Congressional revolt against Executive domination of the weapons procurement process, but the revolt had deep roots going far beyond the objections to any particular weapons system. The essay then assesses the prospects for the future. It analyses the continuing pressures for deployment before identifying some of the considerations which could render ABM controversial once again.

Past Pressures for ABM

Although a history of the ABM project is not feasible in the present context, it is possible to identify those factors which have promoted or facilitated the development and initial deployment of ballistic missile defence.[2] They range from parochial considerations to far-sighted concerns, from narrow assessments of organisational self-interest to genuine and deep-rooted anxieties about the threat to United States

48 ANTIBALLISTIC MISSILE DEFENCE IN THE 1980S

national security posed by Soviet weapons developments, both offensive and defensive. At the most general level, however, is a national approach to strategic problems which extols the virtues of technological solutions and, more often than not, elevates technology into a panacea. The more specific organisational and industrial interests are also important as are those Senators and Congressmen intent on ensuring that the United States maintains a substantial technological lead over the Soviet Union. Although analytically separable, there are, of course, important links between the prevailing attitudes to technology and the organisational infrastructure which has been created to exploit it. As Ted Greenwood has observed, 'Large organisations have been created that owe their continued existence solely to their ability to invent or design new weapons and sell them to political decisionmakers. These organisations include not only the development commands of the services but also some of the largest of the nation's corporations who together employ millions of workers and represent a powerful political force.'[3] In other words, the military and industrial organisations have institutionalised a highly dynamic approach to technology which constantly seeks to advance the frontiers of both knowledge and application.

The attitudes themselves, however, are deeply rooted in American tradition and history. The Research and Development organisations may have strengthened the technological bias in American society and policy but did not create it. The American experience in World War Two, for example, demonstrates very clearly a penchant for technologically intensive warfare rather than a manpower intensive approach. Furthermore, technology has a pervasive and seductive appeal which goes beyond its wartime uses. In the United States, to a greater extent than perhaps anywhere else, highly visible technological achievement is seen as a necessary symbol of a vigorous and dynamic society. Not all observers see this as a virtue of course. Adam Yarmolinsky in particular has been highly critical of what he regards as a short-sighted approach to technological advancement. Describing the American syndrome as a case of 'If you can do it do it,' Yarmolinsky suggests that it has been responsible for numerous over-ambitious development projects.[4] More important it has led to an unbridled technological enthusiasm which pays little regard to consequences – and nowhere has this been more so than in the field of weaponry. Yet this is understandable. The Cold War, not unnaturally, underlined the need for United States technological supremacy. In the late 1940s and early 1950s nuclear superiority was widely regarded as a way of neutralising or compensating for the manpower advantage of the Sino-Soviet bloc. After Sputnik the desire to restore and maintain this superiority was even more urgent. At the same time there was a growing tendency – which was often self-serving in its consequences if not always in intent – to overestimate Soviet strategic capabilities, both offensive and defensive. When faced with the inevitable imponderables about the adversary's military programmes, it is essential that intelligence agencies are prudent in their forecasts. Nevertheless, worst case thinking also has its drawbacks, and when coupled with the desire for superiority has encouraged both 'offset' and imitative' reactions to Soviet weapons developments which were themselves sometimes more apparent than real. The concern in the 1960s over possible Soviet deployment of an extensive ABM system, for example, had a twofold effect. On the one hand, it facilitated the development of MIRV as a way to offset the augmentation of Soviet damage limitation capabilities. On the other, it helped to justify the development and deployment of an American ABM system. If the Russians had such a system, the United States must have one too: 'essential equivalence' of capabilities and options may have been elevated into a formal principle of United States strategic planning by Secretary of Defense Schlesinger in the early 1970s, but it had been an informal guide to action long before then. And because the United States had a technological lead over the Soviet Union in so many areas of weaponry, equivalence was in effect translated into superiority.

Thus a technological bias in US attitudes combined with a legitimate and understandable concern for security to drive the ABM development programme. Although it is important not to denigrate American anxieties, it has to be acknowledged too that Soviet actions were important not only for the dangers they presented but also for the opportunities they offered. Soviet actions were frequently used to justify and legitimise developments and deployments that certain agencies and departments wanted to make for reasons unconnected with national security. This was certainly true of the ABM, which since the mid-1950s has reflected the organisational interests of the Army. It is little exaggeration to suggest that much of the impetus for ABM development arose because the Army was squeezed out of the ICBM programme. After the primary responsibility for IRBMs was also given to the Air Force, BMD was left as the only way the Army could obtain a strategic nuclear programme. As Halperin has noted:

> By the late 1950s the Army realised that its role in strategic nuclear forces would be restricted to defence. Because of the limited role of anti-aircraft defences, those in the Army who were responsible for developing, deploying and operating air defence systems turned to missile defence. For them it was simply the next step in the same direction.[5]

And for the Secretary of Defense, the Nike-Zeus anti-missile missile was a way of compensating the Army for the loss of the Jupiter programme.[6]

Initially, however, the Army was not completely behind ballistic missile defence, with some members suspecting that funding for ABMs would divert resources from other more important roles and missions. Greater unanimity was achieved as a result of budgetary changes instituted by Secretary of Defense McNamara, which meant that ABM would have to compete for resources with other strategic forces rather than be funded out of the Army's budget. Now that concern about the 'opportunity cost' of the ABM was alleviated, the Army could – and did – become much more enthusiastic.[7] Robert McNamara, however, did not share this enthusiasm and in 1963 rejected the Army's proposal for deployment of Nike-Zeus, thereby pushing the ABM 'back into the development stage'.[8]

If the Army had failed to win a victory, however, its defeat was not total. The issue had been kept alive, thereby maintaining the link between the Army and the prime contractor for the ABM, American Telephone and Telegraph (AT&T). It is not necessary to engage in polemics about military-industrial conspiracies to recognise the symbiotic relationship between the military organisations which help to establish weapons requirements and the industrial firms responsible for the research, development and ultimately the production of these systems. The coincidence of interest between the Army and AT&T was apparent as early as 1955 when the Army commissioned a feasibility study of ABM from Bell Laboratories (the research group of AT&T). Although not directly dependent on the ABM contract for its continued viability, AT&T wanted it for reasons of its own. According to Halperin the company 'used its involvement in air defence and missile defence to help prevent an anti-trust suit, to split Bell labs, the research group, and Western Electric, the manufacturing unit, from the Bell system.'[9] Indeed, the company seemed as anxious as the Army to go ahead with an ABM deployment and in 1958 there was a well-coordinated campaign for the upgrading of the system:

> Army, the magazine of the Association of the United States Army, devoted almost an entire issue to the Nike-Zeus. Commanders on active duty contributed ghosted articles praising the anti-ballistic missile. Western Electric and eight sub-contractors bought full page advertisements with maps showing where the $140 million they wanted for Nike would be spent.[10]

The pressure continued during much of the Kennedy Administration and, although it declined after McNamara decided against deployment of the Nike-Zeus, it was to re-emerge dramatically in the ABM controversy of the late 1960s. Once again the Army played a significant role and carried out a major public relations campaign. Drawn up by Secretary of the Army, Stanley Resor, the activities included

Congressional lobbying, contacts with reporters and editors and 'coordination of efforts with contractors in explaining the system.'[11] By then, of course, BMD had become a much more salient issue – and one on which the scientific community was polarised and strategic analysts deeply divided. The defence industry, in contrast, was neither ambivalent nor reticent. This was hardly surprising. In 1969 when the Nixon Administration decided to replace the Sentinel system designed for area defence or city protection with the Safeguard ABM intended to protect ICBM sites, at least 22 companies in 16 states had a major financial stake in continued funding for some form of BMD.[12]

In charge of the development was Western Electric Co, a subsidiary of AT&T. Between July 1, 1967 and April 1, 1969 Western Electric and its subcontractors received more than a billion dollars in connection with the ABM, about two thirds of which was for R & D and the rest for production. The main subcontractors and the amounts paid were as follows:

Raytheon Corp – $109 million for development and production of missile site radars

McDonnell Douglas Corp – $90 million for development and production of the Spartan missile

Martin Marietta Corp – $70 million for the development and production of the Sprint missile

General Electric Co – $45 million for development and production of the long-range perimeter acquisition radars

Texas Instruments Inc – $16 million for production of computer circuit packages

Sperry Rand Corp – $15 million for computer and engineering services and for production of altitude reference units

RCA – $15 million for production of computer circuit packages and for missile range services.

Motorola Corp – $9 million for production of computer circuit packages

Hercules Inc – $8 million for production of Sprint propulsion systems

International Business Machines Corp – $7 million for development of computer programs, computer production and computer rental

Thiokol Chemical Corp – $6 million for production of Spartan propulsion systems.[13]

Given the stakes involved, the lobbying activities of the defence industry seem to have been somewhat muted and restrained. 'However, three major ABM contractors – Motorola, General Electric and Lockheed – contributed financial support to the Chicago based American Security Council, which was lobbying for the ABM system through mailings and broadcasts.'[14] The Council also placed newspaper advertisements and produced 20,000 copies of a booklet arguing the case for deployment. The link between Motorola and the American Security Council was strengthened by the fact that Robert W. Galvin, chairman and chief executive officer of Motorola, was also Chairman of the Council's National Strategy Committee – a subcommittee of which prepared the pro-ABM booklet.[15] There is nothing particularly sinister about such activities. Not only is lobbying a legitimate and indeed necessary feature of the democratic process, but in this particular instance anti-ABM groups were even more vigorous in their presentations. Nor should the 'political clout' of the defence industry be exaggerated: after all, significant restrictions were placed on ABM deployment by the SALT I Agreement. Nevertheless, the pro-ABM lobbying reveals the momentum that had built up behind the system, and even though the 1972 Treaty postponed the deluge of defence dollars widely regarded as a concomitant of a favourable deployment decision, continued research and development contracts may have offered compensation for some although not necessarily all the corporations involved. They also meant that sooner or later ABM would be restored to a prominent place on the defence agenda. As Ted Greenwood has observed,

> The extensive R & D programs that are maintained in order to reduce the risk of technological surprise generate both potential threats and real countermeasures. As a hedge against the realisation of these threats the countermeasures may be actively pursued. At first they would be considered option programs. But if they have important political and strategic advantages, if they gain the support of important sections of the services and the technological community, if they assist in bureaucratic arguments or help maintain the vitality of organisations, and if they are technically both feasible and interesting, they soon generate a large constituency and ensure a life of their own. From then on it may matter very little what the Soviets do, the program is propelled forward by other forces.[16]

The ABM fitted this description almost perfectly – and continues to do so – with the caveat that what Moscow does *is* important, albeit primarily as a legitimising device.

Another part of the constituency for the ABM in the 1950s and 1960s was to be found on Capitol Hill. In the Senate in particular, there were several figures who were prepared to champion the programme, argue the case for greater appropriations and defend it against the critics. Strom Thurmond stood out as probably the most vigorous proponent of ABM, but found strong support from Richard Russell, Henry Jackson and John Stennis, all of whom were solidly conservative and extremely 'reliable' on most defence issues. The fact that consideration of the defence budget in Congress was dominated by the Armed Services and Appropriations Committees provided these pro-ABM Senators with a powerful institutional base from which to operate. The result was that Congress on several occasions appropriated more funds for the ABM than had been requested by the Executive, thereby attempting to speed up or enlarge the programme. Yet it was in Congress that the revolt of the later 1960s against the ABM had its most significant manifestation – and it is to this revolt that attention must now be given.

The ABM Controversy of the Late 1960s

The dispute on ABM which led to a series of extremely close votes on funding was in part the product of the particular circumstances of the late 1960s and early 1970s. Some of the problems, however, were related to the weapon system itself. Doubts about the effectiveness of ABM combined with anxieties about both the costs of deployment and the arms race implications to provoke a reappraisal in a Congress which had hitherto been singularly uncritical in its approach. Had there been unanimity on the technological merits of the system, it is unlikely that this would have occurred.[17] With scientists and technologists deeply divided about the likely efficacy of ABM, however, Congressional opposition seemed far less irresponsible and much more legitimate. Indeed, the anxieties of many Senators mirrored those which had been expressed within the Executive branch itself, most notably by Secretary McNamara. As Alton Frye has observed, McNamara's 'missionary zeal in pointing out the hazards of ABM deployment in the context of the Soviet-American strategic equation . . . won more converts than he expected.'[18] Although McNamara had left the Department of Defense by the time the controversy erupted, his opposition, together with the increasingly public doubts of many scientists, may well have provided the impetus for a Congressional challenge. Just as a united Executive gives Congress little incentive or opportunity to intervene significantly in national security decisions, so divisions within or between departments encourage a more active and probing approach by Congress. Public anxieties have the same effect and it is no coincidence that the Sentinel decision sparked off considerable grass-roots opposition from people who saw ABM deployment as something which would make them a target for, rather than offer protection against, Soviet missiles. It was inevitable that these constituency pressures would be reflected in Congress.

Nonetheless the emergence of a more critical approach was in many ways independent of ABM. One of the reasons the system ran into such difficulty was that the deployment decision coincided with the development of a more sceptical attitude towards military preparedness by many Senators. To a large extent this was an inevitable spillover from the growing disillusionment with the Vietnam War: if the Executive had been so badly wrong in its Indochina policy, it could be equally ill-advised in some of its procurement decisions.[19] Senator Proxmire's revelations about massive cost overruns and the waste and duplication of effort which accompanied much defence spending intensified this disquiet.[20] It may also have contributed to suspicions that deployment of a 'thin' ABM system was merely a 'foot in the door' which presaged a later, and very costly, expansion. In other words, the ABM ran into a 'credibility gap' which was enlarged further when President Nixon's review of the programme resulted in a change of roles and missions which was not accompanied by a comparable change in the weapons themselves. The new Safeguard system was to use many of the same components as the Sentinel even though the military specifications of its task were very different. While this change of rationale made strategic sense, it also suggested that the proponents of ABM were opportunist and far less concerned with the mode of deployment than with the fact of deployment. As a result of the suspicions that were aroused by the switch, the ABM was widely characterised as 'a weapon in search of a mission.'[21]

At the same time, it should not be forgotten that the Senate's protest against ABM was part of a more general revolt against the Presidency. This challenge had both partisan and institutional roots. Certainly the Democratic-controlled Congress was far less inhibited in opposing a Republican President than it had been in its dealings with his Democratic predecessor. There was also a feeling, shared by some Republicans, that the Presidency had become too powerful. Thus the ABM controversy cannot be separated from the assault on the Imperial Presidency: it was one of the early skirmishes in a battle with the Nixon Administration which was soon to extend to the whole gamut of foreign policy and national security issues, from war powers and executive agreements to the American military presence overseas. After years of neglect, deference and acquiescence, Congress was becoming more involved, independent and assertive on national security issues. It was not only less amenable to Executive control, it also was much readier to advance its own preferences. If the ABM helped to catalyse this resurgence, however, it was almost one of the first casualties, and it is difficult to avoid the conclusion that ten or even five years earlier, ABM would have received Congressional endorsement with little difficulty, and certainly with nothing like the close votes which occurred in 1969 and 1970.

The ABM controversy was also symptomatic of a breakdown in the Cold War consensus. The emergence of superpower detente appeared, to many critics at least, to render some of the expenditures on defence superfluous. Such expenditures seemed particularly onerous at a time when the United States was beset by all sorts of internal problems. Thus for many in Congress the ABM came to represent a distorted sense of national priorities in which pressing domestic needs were sacrificed for military programmes which were not only of doubtful reliability but which added little to national security. Liberal Senators in particular protested vigorously against this subordination of national welfare to military preparedness.[22] Just as the ABM debate signalled the beginnings of a Congressional challenge to Presidential authority on national security issues, so it represented the first of a long series of battles in which the liberals attempted to reorder national priorities away from defence. Within the Senate this also involved a revolt against the dominance of the very conservative Committee on Armed Services which was seen as both too deferential and too sympathetic to military requests – a revolt which was led primarily by the senior members of the Committee on Foreign Relations.

Closely related to all this was a changing approach to national security itself. No longer was it simply a case of more being better or of the attainment of security through outspending the Soviet Union. An increasing number of Senators and Representatives were unhappy about a military competition that appeared to have taken on a dynamic of its own and seemed particularly incongruous in a period of political accommodation. Thus mutual restraint was seen as a viable alternative to unrestrained and apparently futile attempts to obtain unilateral advantage. The late 1960s and the first half of the 1970s was a period of unprecedented faith in arms control as something which promised to provide greater security and stability at a lower cost. Indeed, it was the Nixon Administration's acceptance of this point and its pursuit of an agreement limiting ballistic missile defences which effectively defused the ABM controversy in Congress.

By the late 1970s, however, several remarkable changes had occurred. Arms control had fallen into disrepute, as had the notion of Soviet-American detente. And nowhere were the new attitudes more vigorously articulated than in a Senate which had been transformed from the most liberal and 'dovish' institution in the American political system into a conservative and 'hawkish' body intent on forcing an enlarged defence budget on a reluctant President. In this new climate it was not really surprising that the ABM made a comeback. Nor is it surprising that its proponents have learned many lessons from the early controversy and that this time there is a much clearer strategic rationale at the outset. Whether or not this will mean that a decision for ABM deployment is likely to be formulated and implemented without opposition is something which must now be examined.

Current Pressures for ABM

The re-emergence of ballistic missile defence as a live political issue arises, ostensibly, from changes in the perceived strategic relationship of the USA and the USSR, from a reassessment of Soviet 'strategic culture', and from the development of counterforce tactics in both the USA and the USSR. Interest has also been aroused by technological developments which, it is claimed, could provide the USA with an effective 'point defence' of ICBM sites by the late 1980s and a more comprehensive 'layered' BMD system by the early 1990s. In addition to this there has emerged in Congress, and particularly in the Senate, a 'new right' bloc of politicians who have taken an interest in the development of ABM, and who, with the support of defence industrial interests, have pressed vigorously to fund it.

The most dramatic of these changes is the putative vulnerability of American ICBMs to a Soviet first strike, and the allegation that the USSR has acquired, for the first time, a strategic superiority over the USA. The issue of Minuteman vulnerability was first raised in the late 1960s and early 1970s when the USSR began testing its fourth generation ICBMs (SS-17, SS-18 and SS-19), but this concern was muted by and then lost in the growing support for detente. However, it became one of the major issues of the presidential election campaign of 1980, and since Reagan's election has been promoted in one form or another by his Administration. In early 1981, Colin Gray summed up the consensus of the defence community when he wrote, 'there is today no serious argument about the prediction that within a year the US ICBM force, as presently constituted, will be almost totally vulnerable to a Soviet first strike.'[23] On 22 June 1981, Eugene V. Rostow, President Reagan's candidate for Director of the Arms Control and Disarmament Agency, gave evidence to the Senate Foreign Relations Committee and announced that

> The Soviet Union is now close to acquiring a posture from which it could gain an important strategic advantage by striking first or threatening to strike first in a crisis.

He went on to argue that 'If we allow our strategic forces to remain vulnerable to that threat' the USA could be exposed to nuclear blackmail.[24] In July 1981, Edward L. Rowny, Reagan's nominee for chief arms negotiator, argued in similar vein that the threat of Soviet supremacy had become a reality. He claimed that reports from the Secretary of Defense and the Joint Chiefs of Staff showed that 'somewhere between the last quarter of 1980 and the first half of 1981, the Soviets did surpass us in overall strategic superiority.'[25]

It is this state of affairs which President Reagan has described as the 'window of vulnerability', and which Secretary Weinberger claims will be at its widest in the period 1985 to 1986.[26] It has also been claimed that the USSR is continuing to develop its strategic systems, and that a 5th generation ICBM is under development together with the Typhoon class submarine and SS-N-20 SLBMs. Even if the reality of Soviet strategic superiority is questionable, it is argued that the perception of American vulnerability cannot be tolerated, lest the USSR be encouraged and 'emboldened' by it. Colin Gray has emphasised this point and developed it to argue that the over-riding case for the deployment of an ABM system is that it 'will blunt USSR offensive confidence.'[27] These developments and arguments have made the survival of America's counterforce ICBMs a priority and have led to the promotion of a series of passive and active defence solutions for them. As suggested above, with the collapse of a technically viable or politically acceptable deceptive basing mode for existing or new missiles, the case for the active defence of ICBMs using BMD has gained support. It has even been argued that changes in the US/USSR strategic relationship have made 'ballistic missile defense an imperative for the US.'[28]

This so-called 'imperative' has been reinforced by the re-examination of 'Soviet strategic culture' by defence analysts in the USA, and by their conclusion that the USSR not only accepts nuclear war as likely but also believes that it is winnable. In short, the view that holds sway in Washington is that the Soviet Union is equipped for and prepared to fight a nuclear war, and that her strategic modernisation programmes have been designed accordingly. This was apparent in Harold Brown's comments when announcing President Carter's new strategic doctrine in August 1980.

> The Soviet leadership appears to contemplate at least the possibility of a relatively prolonged exchange if a war comes, and in some circles at least, they seem to take seriously the theoretical possibility of victory in such a war.[29]

Under the Reagan Administration, however, the circumspection of Brown's analysis has been replaced by the bald assertion that the USSR subscribes to a nuclear war-winning strategy;[30] an assessment which underpins arguments that the USA must develop and deploy as rapidly as possible a survivable 'hard-kill' missile of its own. Essential equivalence rules.

The reassessment of Soviet strategic culture also suggests that the USSR believes that the deterrent effect of her own weapons derives from their war-fighting capability, and that there is no strategic requirement to attack American cities or to punish American society. In this case 'it would appear that there is no Soviet assured destruction requirement vis-a-vis the American homeland which could be endangered by US area BMD deployment.'[31] Thus there is a clear case for ABM protection of both ICBMs and American cities and, if the assessment of Soviet strategic culture is correct, there would be no destabilising effect on the Soviet-American strategic relationship from such deployments.

The case for counterforce ICBMs defended with ABM also rests upon a shift in US strategic thinking that parallels the alleged emphasis on counterforce in Soviet nuclear strategy. In the early 1970s President Nixon called for an alternative to the

> single option of ordering the mass destruction of enemy civilians in the face of the certainty that it would be followed by the mass slaughter of Americans.[32]

The response to this was National Security Decision Memorandum (NSDM) 242, which became known as the Schlesinger Doctrine. It recognised the possibility that a Soviet attack on the US could take a variety of forms and be less than massive. To counter such threats effectively and to hold out the possibility of terminating a nuclear war before an all-out exchange, Schlesinger argued that the United States must have 'selective nuclear response options' short of assured counter-value destruction. These selective nuclear options amounted to a war-fighting capability for US nuclear forces and were intended first to enhance the chances of terminating a war, but, if that failed, to slow down the recovery of the USSR.[33]

President Carter's 'countervailing' strategy, enshrined in PD59, developed the logic of NSDM 242, but extended its scope. PD59 reflected the view that US deterrent policies should be directed towards those targets which the Soviet political and military leaders valued and should, if not allow the US to win a nuclear war, at least prevent the USSR from doing so. In accordance with this, US ICBMs were targeted against what the Soviet leadership was held to value most – the infrastructure of the regime's power – including the structures of political and military command and control. Consequently the target list grew from 25,000 in 1974 to over 40,000 in 1981.[34] A war-fighting capability remains a high priority in the current US strategic warfare philosophy. Lt. General James W. Stansbury has said in evidence to Congress that in a nuclear war the US must be prepared to fight and go on fighting, and that 'an eight-hour nuclear war is no longer an acceptable concept.'[35] These strategic requirements make the survival of a counterforce ICBM a high priority, and it was this that made the development of a secure basing mode for the MX missile such an important issue for both the Carter and Reagan Administrations, thereby facilitating a resurgence of interest in ABM. In contrast with the 1960s, the proponents of ABM have now developed a clear-cut and more compelling strategic rationale. Indeed, Soviet advances in technology, whether real or fabricated, are being highlighted to legitimise an augmented ABM programme. This time round, therefore, it will be more difficult – although as discussed below, not impossible – for critics of ABM to characterize ABM as a weapon in search of a mission.

Another advantage of the proponents of BMD is that the technology of strategic defence has improved considerably over the last decade. Indeed, the frequency with which BMD solutions are offered for point defence of missiles is explained in part by the view that the technology now exists to make these systems highly effective. Clarence Robinson has recently written that 'there is general agreement that the US is not technology limited, only funding limited.'[36] Yet, much of the apparent increase in the efficiency of new programmes is the result of a reduction in their operational requirements compared to the earlier Sentinel system. When President Nixon changed the role of BMD on 14 March 1969 by substituting point defence of Minuteman silos for the 'thin' area defence expected of the Sentinel system, he implicitly accepted 'leakage', or a less than perfect defence. Since then the most developed ABM programmes also employ 'preferential defence' which requires the system to defend only those silos which still contain an active missile.

At the same time there has been genuine improvement in ABM technology as a result of work carried out by the Army's Ballistic Missile Defense System Command. The Army controls two programmes the first of which, the Systems Technology Program (STP), is mainly concerned to develop and demonstrate the Low Altitude Defense System (LoADS) for point defence of missiles by the mid-1980s. It was the Army's belief that LoADS would employ low-risk and 'mature' technology, such as the General Electric phased array-radar and TWR Systems software, that persuaded the DOD to fund and authorise a pre-prototype demonstration for the mid 1980s.[37] The programme also includes the Homing Overlay Experiment (HOE) for the development of an exo-atmospheric BMD system to operate 'synergistically' with LoADS as the upper level of a layered BMD system. HOE is designed to carry a non-nuclear 'kill' device, on-board sensors, and will work in conjunction with an independent probe which will identify incoming warheads and guide the interceptors to them. The Army has announced that it believes that HOE can become effective rapidly as a result of techniques already developed and demonstrated in its second programme, the Advanced Technology Program (ATP).[38]

The main purpose of the ATP is to examine long-range technologies, especially radar and on-board sensors, for future generations of BMD. Indeed, considerable effort in both the STP and ATP will go into the development of an exoatmospheric interceptor, and this reflects the decision in October 1979 to extend the 'baseline' of the ABM programme from point to layered defence. A layered defence system will not become an operational possibility before the mid-1990s, but there is massive confidence that the LoADS system could be operational very quickly. Maj. General Grayson D. Tate Jr., the Army's BMD programme manager, has said that tests had

> proved beyond a reasonable doubt that we have the technology to build an effective terminal defence system that can detect, discriminate and intercept ICBM warheads, even in the extreme environment caused by massive ICBM attacks, ICBM tank fragments and penetration aids.[39]

Such confidence has been boosted by other technological developments that hold out the possibility of building 'autonomous interceptors' which would use on-board guidance rather than ground-based command guidance. The interceptors would use very high speed integrated circuits, on-board sensor, real-time processing and logic capabilities that 'border on artificial intelligence,' and would work in conjunction with sensor probes which would free them from the very vulnerable and easily confused ground based radars.

In addition to these programmes, private companies have funded several others that are designed for endoatmospheric and exoatmospheric BMD. Vought are developing the Quick Shot programme which uses a barrage of small missiles in a controlled 'shotgun blast' for point defence. A similar programme is under development by Tracor's MB Associates Division under the name of Swarmjet, but this and Quick Shot suffer from the turbulence created by the simultaneous launch of many small missiles. A rather more novel approach to point defence is being undertaken by Sandia National Laboratory who are also looking at the use of small missiles, but in this case designed to saturate the re-entry corridor with steel cubes, and which could use relatively cheap radar. Space-based systems were also examined recently in the 'High Frontier' study, and the advantage of such systems is that they could intercept ICBMs in the boost phase, thereby avoiding the collateral damage to the defending country normally associated with endoatmospheric defence.

Such systems, however, are likely to depend upon the use of lasers and particle beams, and it is accepted that operational technologies of these kinds are 10-15 years and 20 years away respectively.[40]

It is clear though that the links established in the 1950s and 1960s between the Army and some of the major industrial contractors remain strong, and have provided much of the impetus for the continued development of BMD technology. And once again considerable support has come from Congress, particularly from the Armed Services Committees of both Houses, backed by a powerful coalition of conservatives. Senators Harrison Schmitt, Malcolm Wallop, Pete V. Domenici, Jake Garn and Paul Laxalt have emerged as the spokesmen of the 'new right' in defence affairs, and have been very influential in pressing Congress to accept the ABM. The direct successor of the pro-ABM Senators of the 1950s and 1960s, this group is sufficiently powerful to threaten to 'legislate military strategy' even if the President has second thoughts about missile defences.[41] Their strength stems from the widespread concern in Congress about America's strategic position vis a vis the USSR, and the responsiveness that most Congressmen have to a public opinion which, until recently, unequivocally supported President Reagan's pledge to 'make America strong again.'

This congressional and public mood highlights the success of lobby groups such as the Heritage Foundation and the American Conservative Union, and more particularly the defence groups such as the Committee on the Present Danger and the American Security Council, all of whom have been weighty opponents of detente and persuasive advocates of increased defence spending and of strategic modernisation including ABM.[42] It is no secret that many of these groups, including the American Security Council, are assisted with information and funds by the defence Contractors' Political Action Committee, and it is also known that the average annual contribution of aerospace companies to that fund is $81,000.[43]

The strength of the defence industrial lobby for BMD reflects in part the large numbers of contractors that are already involved in its development, and the vast amounts of money that may be expended on developing these systems. This last factor is of vital importance to Congressmen in so far as it represents potential investment and employment in their congressional districts and in their states. The current expenditure on ABM is small compared to the over-all size of the defence budget, but the DOD plans to spend up to $8.5 billion on various systems by Fiscal Year 1987, and within this period it is estimated that the LoADS programme will cost £1.4 billion. It must also be remembered that BMD is likely to be part of a wider strategic modernisation programme, and Lawrence Korb has estimated that an MX programme involving ABM will cost $80 billion, while the Congressional Budget Office has put the price at $96 billion.[44]

As in the past, the ABM programme involves some of the major defence contracting firms. McDonnell Douglas is the prime operator of the Army's BMD Systems Technology Programme, and will also act as the systems integrator of the LoADS programme which will be based on a one-stage interceptor built by Martin Marietta-Orlando, and a phased-array radar built by General Electric. The radar itself uses a Control Data Systems computer and TRW systems software. The prime contractor for the HOE programme is Lockheed Missile and Space Co., and contracts for sensor development have been given to Honeywell, Boeing, Hughes Aircraft and Rockwell International. Computer developments for the programme have gone to Lockheed Research Centre, and radar development has been given to Lincoln Laboratory which is part of MIT. Other radar contracts for equipment to be used on the Kwajalein Test Range have gone to RCA, Tradex and GTE Sylvania.[45]

Against this background it is hardly surprising that the Reagan Administration made a commitment to ABM in its five point strategic modernisation programme announced in Ocober 1981, although the strength of that commitment is not unqualified. The key statement in the programme notes that

> Any ground-based scheme (for ICBMs) ultimately would require a ballistic missile defence for survivability. But, today, ballistic missile defence technology is not at the stage where it could provide an adequate defence against Soviet missiles. For the future, we are not as yet sure how well ballistic missile defenses will work; what they will cost; how Soviet ballistic missile defence – which would almost certainly be deployed in response to any US missile defense system – would affect US and Allied missile capabilities; and what would be the political ramifications of altering the ABM Treaty.

This statement has been interpreted quite differently; the IISS sees it as indicating a cautious approach by the Reagan Administration, whilst Senator Pete V. Domenici regards the statement as a 'mandate to remove the wraps from ballistic missile defense research and development and examine it as a serious force option.'[46] Whatever the precise intent of the Reagan Administration, the statement shows that it is alive to, and indeed may anticipate, the many objections that could be raised against a large investment in BMD, despite its many supporters. Even so, in his report to Congress for Fiscal Year 1983, delivered in February 1982, Caspar Weinberger restated the Administration's policy of seeking a margin of strategic safety and stated that our goal will be to gain and maintain a nuclear deterrent force which provides us with an adequate margin of safety with emphasis on enduring survivability . . .[47]

Such an emphasis is likely to make BMD a continuing consideration for the Reagan Administration. Yet there are some signs that BMD may once again run into opposition. Although a repetition of the full scale Congressional battles of the late 1960s is unlikely, there are several factors that may qualify or inhibit the prospects for BMD in the coming years.

Potential Opposition to the ABM

The revival of interest in ABM has led, almost inevitably, to a re-examination of America's strategic philosophy, and in particular of the claim that US ICBMs are vulnerable to a Soviet first strike. There is a growing body of opinion which holds that such a threat does not exist and, even if it does, that the utility of land based missiles may well be declining and with it the relevance of ABM point defence. Doubts have also been raised about the ability of either the USA or the USSR to develop a reliable counterforce capability, and fears have been expressed about recent shifts by the United States towards a war-fighting strategic philosophy. In addition, it is clear that major technological advances notwithstanding, there are still considerable technological problems to be overcome and that any form of ABM deployment is unlikely before the early 1990s.

The strategic and technological problems may not be the most difficult for the supporters of BMD when compared to the economic and political difficulties. The willingness of Congress and the American public to pay for a major expansion in defence capability, including ABM, depends in large degree upon the success of Reaganomics, yet it is already clear that the Administration has been unable to cope with many of the major problems that afflict the American economy. Indeed, there is a growing body of opinion which believes that the size, pace and content of Reagan's defence build-up compound those economic problems. Furthermore, there has been a powerful public reaction to the 'sabre rattling' that has accompanied Reagan's defence build-up. Although most marked in Europe where the CND and END movements command considerable support, there is also a new and influential CND-like movement developing in the USA, and considerable support in Congress for a nuclear freeze. The effect has already been to accelerate President Reagan's plans for arms control negotiations with the USSR, and to de-emphasise the strategic modernisation programme. If such pressures continue it may be politically impossible to upgrade a programme as potentially contentious as ABM. The more visible and costly it becomes, then the more likely it is to generate opposition.

This is all the more likely as the strategic rationale of BMD is called into question. And despite the vigorous and unequivocal arguments of the system's proponents, this cannot be ruled out. Indeed, there are growing doubts about the claim that US ICBMs are vulnerable to a Soviet first strike. At present, even assuming no degradation of accuracy from test range values, the USSR would have to target at least 2,000 warheads against the more than 1,000 US ICBM launchers in order to achieve a 90 percent or so level of destruction. It would have to launch these 2,000 warheads in perfect coordination and, in the view of the IISS, 'the command and control requirements of such coordination border on the infeasible.'[48]

The second, and more substantial criticism of the view that US ICBMs are vulnerable centres around the accuracy of modern missiles. Prominent members of the Townes Committee, set up by President Reagan to re-examine the vulnerability of ICBMs and the alternative basing modes for the MX missile, have pointed out that the effect of gravitational bias, the turbulence caused by a salvo launch, the inability to test over war-time flight paths and the problem of 'fratricide' (in which the first warheads to land either pre-detonate or deflect the remaining warheads) all make for an accuracy far less than the 600 ft CEP usually ascribed to Soviet missiles. Professor J. Edward Anderson of the committee has said that 'ICBMs are and will remain city busters, not silo busters.'[49] The Administration itself seems to have revised its view of ICBM vulnerability, and the Department of Defense's annual report for FY 1982 notes that, given warning, the US would lead the USSR in 'survivable warheads'. Warner R. Schilling has also pointed out that 'the second strike capability of the US was larger in 1980 than it was in 1964, when the Soviet build-up began' provided that such capability is measured by the number of nuclear weapons that would survive an initial exchange.[50] After the Reagan Administration decided on an interim basing of MX missiles in Minuteman silos, USAF officers testified to Congress that MX survivability is only an issue in the event of an all-out first strike by the USSR, and that such a strike is unlikely. Even if the survival of US land based missiles was at risk, Gregory Treverton has pointed out that this would be a problem for deterrence in Europe but not between the superpowers.

> Whatever the elaborate analytic constructs the strategic theorists create, it is hard to see that ICBM vulnerability threatens the US's ability to deter Soviet nuclear attacks on herself, even ones directed solely against ICBMs.

The reason for this is that a Soviet attack on ICBMs would cause massive devastation and from 5 to more than 20 million dead, and no American President would fail to reply with any remaining ICBMs and with SLBMs. Treverton goes on to argue that 'in that sense the super-power deterrence is quite robust.'[51] If this view gains ground, one of the major justifications for ABM will be removed.

There are also many technological problems that still stand in the way of an effective BMD system. Even the most promising short-term programme, LoADS, must be further developed to improve the 'hardness' of its radars, to ensure that they work after long periods of 'dormancy' and to improve the engagement time of a system that will have less than 10 seconds to intercept a warhead at below 50,000 feet.[52] Perhaps more important than this is the fact that there appears to be no effective way to make ground launched ICBMs invulnerable. The massive study of MX basing options carried out by the Office of Technology Assessment concludes that there can be no basing mode for MX that would solve the vulnerability problem 'much before the end of the decade.' In October 1981 Richard Perle, the Assistant Secretary of Defense for International Security Policy, said that 'there is no conclusive alternative today for a single survivable land-based force,' and that a combination of bombers, Trident D5 SLBMs and cruise missiles 'may aggregate into a strategic balance.'[53] His proposals included the afterthought that such a balance may include a deep underground basing mode for the MX with some element of active defence, but recent studies have shown that the prospects of a viable and acceptable deep basing mode are poorer than those for the aborted multiple protective shelter system.[54] If no satisfactory basing mode can be found for MX and other land based missiles, the need for BMD is reduced. This together with the fact that no ABM system will be operational before the 1990s has led the IISS to conclude that:

> There is ample reason to believe that the mission for which ballistic missile defences are deemed most suitable – defence of point targets – is losing its importance and will continue to do so. The growth of offensive inventories, deficiencies in . . . BMD technologies, and new strategic concerns will contribute to the declining significance or manageability of the once-central issue of ICBM vulnerability. The BMD concepts or hardware now being given serious consideration, will directly affect few of the strategic concerns of the future: vulnerability of command and control facilities . . . , strategic applications of cruise missiles and potential countermeasures, and the impact of anti-satellite or space systems.[55]

Recent events in Congress have reflected these doubts and the strong support for President Reagan's first defence programme has been replaced by scepticism about his second. There are growing demands from both parties that his defence budget be reduced. Congress has cut $2.2 billion from the MX programme until such times as the Administration is able to devise an effective and politically acceptable basing mode,[56] and this will delay not only its deployment but also the development of its associated BMD programmes. Indeed, in May 1982 DOD officials feared that the MX programme might be terminated before approval could be gained for a basing mode acceptable to Congress.[57] If this should occur, the major argument for the ABM would be removed. In any event, pressure to reduce the military budget will cut BMD funds, and in a package of $7.2 billion defence cuts offered to Congress in May this year, Weinberger proposed that $200 million be taken from the Army's STP BMD programme. This represents the first reduction for several years in BMD provision.[58] It may not be the last.

Indeed there is likely to be sustained pressure in Congress for defence cuts so long as President Reagan retains the broad outlines of his supply-side economic strategy, and so long as the American economy continues to respond badly to it. The pressure has become more intense with the publication of several studies that purport to show that the defence build-up itself has caused or has made worse many of America's economic problems. Lester Thurow, a professor of economics at MIT, and Henry Kaufman of the Wall Street investment firm Salomon Brothers, have emerged as the major critics of Reagan's defence programme. Thurow has estimated that the current five year defence programme represents a military build-up three times greater than that of the Vietnam war, and he, along with Kaufman, has claimed that it will inevitably drive up inflation and reduce America's competitiveness against Europe and Japan.[59]

The problem stems from the lack of skilled labour and technical talent in the American economy; as arms industries increase their demand for skilled labour so its price will rise and this in turn will drive up inflation. The demand for technical talent will also deny the civil economy those skills which have given American industry a competitive edge in world markets, and even more ground will be conceded to European and Japanese producers who already have a higher investment of these skills in their civilian economies than do the Americans. These arguments have been countered by Murray L. Weidenbaum, the Chairman of the Council of Economic Advisers, who claims that the military build-up will be smaller and slower than that for Vietnam, and that the spare capacity in the American economy will be able to absorb the increase without an adverse effect on inflation. However, even the most cautious of economists now agree that the scale of Reagan's defence build-up in the context of America's current economic problems is likely to undermine his economic strategy and could do permanent damage to the American economy. Charles L. Schultze has argued that

> there is nothing inherently inflationary or productivity lowering about defence spending that should prevent the United States from having the level of spending it believes is required for its national security, *so long as it is willing to pay for increased defence through lower spending on consumption.*[60]

Schultze has also suggested that the heavy emphasis on procurement in the Reagan build-up will have an inflationary effect on that sector of American industry which produces for the DOD and this will lead to a higher rate of inflation for military goods than for civilian goods, but that this will not have a macro-economic effect if consumption is reduced either by public expenditure cuts in welfare programmes or by increased taxation. President Reagan, however, has cut welfare expenditure but has cut taxation as well, and it is clear that this combination is not sufficient to pay for the vast increase in military spending. The result has been an increased deficit. Indeed the Office of Management and the Budget has recently estimated that it could be as high as $182 billion for the current fiscal year.[61]

Against this background, Congressional attention has switched from the needs of defence to those of the economy, focusing on the urgent need to reduce the deficit. Congressman Les Aspin captured this mood when he said recently that

> The Soviet Union is still in Afghanistan and the military balance hasn't appreciably changed. What changed is that everybody is worried about the deficit.[62]

Furthermore, there is widespread agreement that the only way that the deficit can be reduced in the present political circumstances is to cut defence spending. So strong is this feeling that some of the most powerful supporters of Reagan's military policies have spoken out in favour of defence cuts, justifying them in terms of the need to protect the economy. Senator D. Durenburger has called for a $3 billion cut in the current year and a $25.9 billion cut over five years on the grounds that the economy cannot stand the rate of increase called for by Reagan. Representative Lee H. Hamilton, a sub-committee chairman of the Joint Economic Committee has said:

> There is substantial evidence to conclude that, as presently
> planned, the defence buildup will contribute to the widening
> Federal deficit and to industrial bottlenecks and inflation. It is also
> possible that the buildup will worsen the problem of cost overruns
> in defence procurement.

He has therefore called for a considerable slowdown in the rate of
military spending. Other, normally strong defenders of defence
spending, such as the ranking Democrat on the Senate Budget
Committee Ernest Hollings, have called for a freeze of military spending
at the FY 1982 level, and outright opponents have taken the opportunity
to criticise Reagan's military budget, as when Senator Carl Levin described it as 'a declaration of economic war on America.'[63]

Such criticisms and demands for spending cuts have not been directed
specifically at BMD programmes, but these are certainly not immune.
Congress is now more alive to the strategic, technological and cost
question-marks that hang over the ABM, and its development may suffer
as a consequence of congressional concerns about the MX missile. Still in
the research and development phase, the ABM programme is vulnerable
to a reduction in funding with a consequent loss of momentum.

A final set of pressures that may work to the disadvantage of the
supporters of BMD arise from public reaction to Reagan's 'sabre
rattling' and the decision in the early part of his Administration not to
negotiate arms control with the USSR. It has been further compounded
by the size of the US military buildup, by the promotion of a nuclear
war-fighting strategy, by the Administration's demands that NATO
must modernise its theatre nuclear weapons, and by the President's
unguarded statements about the possibility of limited nuclear war in
Europe. These have served to promote a 'warmonger' image for Reagan
and a widespread belief in Europe and America that his policies represent
a greater danger to world peace than do those of the USSR. It is also
clear that Reagan's vigorous promotion of defence, his harsh criticisms
of the USSR and his refusal to entertain their early initiatives for arms
control have reinforced Soviet peace propaganda that has been aimed at
the anti-nuclear groups in Europe.

The political influence of those groups was reflected by Reagan's
sudden change of rhetoric in late 1981. In January that year he had
described the USSR as a state that would 'reserve unto itself the right to
commit any crime, to lie [and] to cheat,' and declared that there was little
prospect of arms negotiations between the super-powers. Yet, in a major
foreign policy speech on November 18th he announced that the US
would go to Geneva to discuss the modernisation of theatre nuclear
weapons, and that America would 'negotiate in good faith and [was]

willing to listen to and consider the proposals of our Soviet counterparts.'[64] Much the same pressures were responsible for the decision to send Secretary of State Haig to Geneva in January 1982 to discuss with Gromyko the opening of the Strategic Arms Reduction Talks (START) which, although delayed by the Polish crisis, eventually began on 30th June 1982. These pressures have been reinforced by the growth in the US of several anti-nuclear groups on a scale not seen before, and there is also a growing number of Congressmen who are supporting the proposals of Senators Edward Kennedy and Albert Gore for a nuclear freeze at current levels of deployment.[65]

These groups have attacked the strategic philosophy of the Reagan Administration and the strategic modernisation programme itself. On both counts BMD has come in for substantial criticism. In so far as BMD is deemed vital for a counterforce strategy, it has come in for particular criticism, and it is also denigrated as one of the five major elements of the modernisation programme. These criticisms are likely to continue to weigh heavily with the Reagan Administration, and may impose a severe constraint on any desire to renegotiate and relax the 1972 ABM Treaty in such a way as to allow the full-scale deployment of BMD. Reagan may well have to accept the Treaty in its present form at the review conference to be held later this year, including its restriction on the widespread deployment of BMD, since it is the only surviving symbol of arms control and there is much support for it in Europe and America.

Conclusion

The implication of all this is that the immediate outlook for the ABM is far less favourable than its proponents may have hoped. Indeed, there are several features of the current political climate which—rather surprisingly in view of the trends of the 1970s—are reminiscent of the late 1960s. In particular, there are extraneous considerations which make an enlarged BMD programme both less attractive and less likely than might have been expected. Once again there is widespread concern over national priorities. In this case, however, it is not simply that the high level of defence spending appears onerous given the sacrifices being made in other areas such as social welfare; equally, if not more, important is the belief that such spending, and the accompanying budget deficit, is damaging to the economy. Another, and even less predictable, similarity with the earlier period is that there is again some hope for moderating and regulating the arms race through formal negotiations with the Soviet Union. How strong the Administration's commitment to START is remains to be seen. Nevertheless, it is a fact of American political life that it is far easier for a hard-line President to obtain politically acceptable arms control agreements than it is for a more liberal Administration. Consequently, the initial expectations of ABM proponents that the Reagan Administration would be sympathetic to their cause may prove to be illusory.

Indeed, far from expanding the development programme as hoped, BMD supporters may become increasingly preoccupied with a holding operation. With growing pressure on the defence budget, damage limitation, in the political sense, may be the best that can be hoped for. Furthermore, the next few years could be a particularly inopportune time for upgrading the programme: the more salient and expensive ABM becomes, the more controversial it is likely to be. At the same time, it should not be anticipated that BMD will disappear from the strategic agenda; the pressures which have sustained it since the 1950s will almost certainly prevent that. In doing so, they will keep the option open for an eventual expansion of the programme and a large-scale deployment of BMD. For the moment though, the domestic political context in the United States suggests that the ABM Treaty of 1972 is not in immediate danger.

NOTES

1. W. S. Hyland, 'The Soviet View' in 'ABM Revisited: Promise or Peril?', *Washington Quarterly* (Autumn 1981), p. 67.
2. A detailed overview of the ABM development can be found in J. Hadley, *The Anti-Ballistic Missile Debate: the Politics of a Weapon Procurement* (dissertation presented for the M Litt in Strategic Studies, University of Aberdeen 1981). This work was particularly useful in the preparation of the historical part of this paper.
3. T. Greenwood, *Making the MIRV: A Study of Defence Decision Making* (Cambridge, Mass.: Ballinger, 1975), p. 13.
4. See Yarmolinsky's contribution to 'ABM Revisited: Promise or Peril?', op. cit., p. 79.
5. M. Halperin, 'The Decision to Deploy the ABM: Bureaucratic and Domestic Politics in the Johnson Administration', *World Politics,* Vol. 25, No. 1 (Oct. 1972), pp. 62-95 at p. 67.
6. C. Murdock, *Defence Policy Formation: A Comparative Analysis of the McNamara Era* (New York: Suny Press, 1974), p. 117.
7. Halperin, *op. cit.*, pp. 67-8.
8. McNamara's position is discussed most fully in Murdock, *op. cit.*, pp. 116-37.
9. Halperin, *op. cit.*, p. 70, footnote 8.
10. J. Goulden & M. Singer, 'AT&T and the ABM' in C. W. Pursell (ed.), *The Military-Industrial Complex* (New York: Harper & Row, 1972), p. 243 quoted in Hadley, *op. cit.*, p. 17.
11. 'Public and Private Lobbying Involved in ABM Fight', *Congressional Quarterly Almanac 1969* (Washington: Congressional Quarterly Inc., 1970), pp. 1090-6 at p. 1091.
12. *Ibid*, p. 1092.
13. See *ibid*, p. 1092.
14. *Ibid*, p. 1092.
15. *Ibid*, p. 1096.
16. Greenwood, *op. cit.*, p. 149.

17. A. Frye, *A Responsible Congress: the Politics of National Security* (London: McGraw-Hill, 1975), p. 29.
18. *Ibid*, p. 22.
19. This point is made in *ibid*, p. 21 as well as by Greenwood, *op. cit.*, p. 117.
20. For a full account of his findings see W. Proxmire, *Report from Wasteland* (New York: Praeger, 1970).
21. The description is that of Senator Albert Gore and is quoted in Frye, *op. cit.*, p. 31.
22. That many of the debates of this period revolved around the question of whether priority should be given to 'national security' or 'national welfare' is persuasively argued in R. Z. George, *Contending Views of National Security*, PhD thesis (Boston: Fletcher School of Law and Diplomacy, 1977).
23. Colin S. Gray, 'A New Debate on Ballistic Missile Defence', *Survival*, Vol. XXIII, No. 2, March/April 1981, p. 64.
24. See report of Hearings in Michael R. Gordon, 'A Matter of Perception', *The National Journal*, Vol. 13, No. 27 (4 July, 1981), p. 1215.
25. See Paul Mann, 'US to Seek Arms Reduction, Amendment to Treaty on ABM', *Aviation Week & Space Technology* (27 July, 1981), p. 60.
26. Caspar Weinberger, *Testimony to the Senate Armed Forces Committee, 5th October, 1981*. Extracts reprinted in *Survival*, Vol. XXIV, No. 1 (Jan./Feb. 1982), pp. 29-31.
27. Gray, *op. cit.*, p. 68.
28. Clarence A. Robinson, 'Emphasis Grows on Nuclear Defence', *Aviation Week & Space Technology* (8 March, 1982), p.27.
29. *Address by US Secretary of Defense Harold Brown, 20th August, 1980*. Reprinted in *Survival*, Vol. XXII, No. 6 (Nov./Dec. 1980), pp. 267-9.
30. See for example Richard Pipes, 'Why the Soviet Union Thinks it Could Fight and Win a Nuclear War,' *Commentary*, Vol. 64, No. 1 (July 1977), pp. 21-34, and Colin Gray, 'Nuclear Strategy: The Case for a Theory of Victory,' *International Security*, Vol. 4, No. 1, pp. 54-87.
31. Gray, *op. cit.*, p. 65.
32. Richard Nixon, *A Report to Congress: US Foreign Policy for the 1970s—A New Strategy for Peace* (18 Feb., 1970), p. 122.
33. See Leon Goure, 'The US "Countervailing Strategy" in Soviet Perception', *Strategic Review*, Vol. IX, No. 4 (Fall 1981), p. 53.
34. *Ibid*, p. 58.
35. Reported in Clarence A. Robinson, 'Pentagon Backs Strategic Modernisation', *Aviation Week & Space Technology* (26 Oct., 1981), p. 53.
36. Robinson, 'Emphasis Grows on Nuclear Defense', *op. cit.*, p. 33.
37. 'Special Report: Demonstration Planned for MX Defense System', *Aviation Week & Space Technology* (16 June, 1980), pp. 220-8.
38. Philip J. Klass, 'Ballistic Missile Defense Tests Set', *Aviation Week & Space Technology* (16 June, 1980), pp. 213-18.
39. Quoted in 'Special Report', *op. cit.*, p. 228.
40. *Strategic Survey 1981-1982* (International Institute for Strategic Studies, London 1982), p. 15.
41. William Bader, 'Congress and the Making of US Security Policies' in 'America's Security in the 1980s: Part I', *Adelphi Papers*, No. 173 (Spring 1982), pp. 17-18.
42. See Stephen Kirby, 'Congress and National Security', *The World Today*, Vol. 37, Nos. 7-8 (July/Aug. 1981), pp. 270-7, *passim*.
43. Michael R. Gordon, 'Are Military Contractors Part of the Problem or Part of the Solution', *The National Journal*, Vol. 13, No. 28 (11 July, 1981), p. 2133.
44. For a discussion of the cost of the various MX/BMD options see Michael R. Gordon, 'For Reagan the MX Missile may represent "the Strategic Decision of the Decade" ', *The National Journal*, Vol. 13, No. 7 (14 Feb., 1981), pp. 260-4.
45. See 'Special Report', *op cit., passim*.
46. Pete V. Domenici, 'Toward a Decision on Ballistic Missile Defence', *Strategic Review*, Vol. X, No. 1 (Winter 1982), p. 23, and *Strategic Survey 1981-1982, op. cit.*, pp. 15-16.

47. *Secretary of Defense Weinberger's Report to the Congress for the Fiscal Year 1983* (8 Feb., 1982). Extracts reprinted in *Survival*, Vol. XXIV, No. 3 (May/June 1982), pp. 131-4.
48. *Strategic Survey 1979-1980* (International Institute for Strategic Studies, London, 1980), pp. 14-15.
49. See Paul S. Mann, 'Panel Re-examines ICBM Vulnerability', *Aviation Week and Space Technology* (13 July, 1981), pp. 141-8. Anderson quoted on p. 141.
50. Warner R. Schilling, 'US Strategic Nuclear Concepts in the 1970s', *International Security*, Vol. 6, No. 2 (Fall 1981), pp. 48-79.
51. Gregory Treverton, 'Nuclear Weapons in Europe', *Adelphi Papers*, No. 168 (Summer 1981), p. 3.
52. 'Special Report', *op. cit.*, p. 220.
53. Quoted in Clarence A. Robinson, 'Decision Reached on Nuclear Weapons', *Aviation Week & Space Technology* (12 Oct., 1981), p. 23.
54. See Perrin Clausen, 'Study Finds MX Deep Basing Problems', *Aviation Week & Space Technology* (17 May, 1982), pp. 191-2.
55. *Strategic Survey 1981-1982, op. cit.*, p. 18.
56. For a report of the congressional debate see Clarence A. Robinson, 'MX Production Runs into Opposition', *Aviation Week & Space Technology* (5 April, 1982), pp. 22-3.
57. See Clarence A. Robinson, 'Administration Refines MX Basing Plan', *Aviation Week & Space Technology* (3 May, 1982), pp. 14-15.
58. See Clarence A. Robinson, 'Defense Secretary Offers List of Cuts to Congress', *Aviation Week & Space Technology*, (17 May, 1982), pp. 18-20.
59. Lester Thurow, 'How to Wreck the Economy', *New York Review of Books*, Vol. XXVII, No. 8 (6 April, 1981), pp. 3-6, and Michael R. Gordon, 'If Defense Spending is on the Rise, Can Inflation be Very Far Behind?', *The National Journal*, Vol. 13, No. 25 (20 June, 1981), pp. 1101-5.
60. Charles L. Schultze, 'Economic Effects of the Defense Budget', *The Brookings Bulletin*, Vol. 18, No. 2 (Fall 1982), p. 2.
61. Reported in 'Defense Cut of $5 Billion Expected in Budget Action', *Aviation Week & Space Technology* (7 June, 1982), p. 21.
62. Quoted in Michael R. Gordon, 'For the Pentagon's "Minimal Budget" It Is Not Whether to Cut, But How Much', *The National Journal*, Vol. 14, No. 13 (27 March, 1982), p. 544.
63. Hamilton and Levin are quoted in Alton K. Marsh, 'Senate Defense Cuts Emerge', *Aviation Week & Space Technology* (22 Feb., 1982), p. 2.
64. Ronald Reagan, *US Foreign Policy*, 18 Nov., 1981. For a discussion of this and other statements by the Reagan Administration see Dick Kirschten, 'Sheathed Sabers', *The National Journal*, Vol. 13, No. 48 (28 Nov., 1981), p. 2128.
65. For a review of the anti-nuclear movement in the USA, including congressional support for a nuclear freeze, see 'Special Report: The Nuclear Nightmare', *Newsweek* (26 April, 1982), pp. 20-9.

Europe and the ABM Revival

Lawrence Freedman

West European members of NATO have a number of possible interests in the revival of activity in the field of anti-ballistic missiles (ABMs). These interests can be divided into two: those that might conceivably flow from the maximum possible objectives behind the current research effort being realised, and those that are likely to flow from the super-powers simply making the effort.

In the first category it is necessary to consider the consequences for Europe of the general upheaval in all strategic assumptions that would follow from a successful breakthrough in ABMs by either the United States, or the Soviet Union, or both together. In a sense, ABMs represent the last best hope of a real strategic superiority. It is a measure of the desperation of those seeking such a superiority that they are still attempting to revive an option that requires enormous economic and scientific investment, that is severely restricted by international treaty and that has always suffered in practice from the inherent advantages accruing to the offence in nuclear warfare.

A number of American strategists are now arguing that the only way to save the US nuclear guarantee to Europe would be to neutralise Soviet offensive strength by an effective defence. Alternatively a Soviet success, it is argued, would completely undermine the guarantee because the US would be left so vulnerable. If the two super-powers are successful more-or-less simultaneously, then the result will be to re-order the strategic balance on a basis that is somewhat less terrifying but possibly more uncertain than that involved in mutual assured destruction.

European attitudes towards all these possibilities are likely to be shaped first by whether they will be found in front of or behind the defensive barrier to be constructed by the United States. If Europe were to be protected as much as the United States then that would be a cause for general satisfaction. Despite some limited research on anti-tactical ballistic missiles (ATBM) Europe cannot confidently expect such protection and so its position would be extremely uncomfortable for it would now offer the most available set of Western targets for the Soviet Union, and could thus find itself as a sort of hostage.

When we turn to the second category of issues, this very possibility of Europe being left alone exposed, between the trenches as it were, could well have an impact on views in Europe as to its long-term strategic position and the value of the relationship with the United States. Even as an indication of a US aspiration to superiority, a visible and determined ABM development effort in the US could unsettle opinion in Europe and lead to questions as to the wisdom of associating with a super-power which was intent on such a high-risk strategy. Questions of this sort are also likely to result from any abrogation or even just amendment of the 1972 ABM Treaty in order to pave the way for the full-scale exploitation of the new defensive technology. As this would be taken to undermine the cause of arms control such a move would be regretted in Europe, given the general disposition in favour of this cause. Even if the hypothetical gains of effective defence were accepted, an immediate cost of unravelling the whole arms control process, which is politically extremely important, would not be considered worth paying. Lastly, in this category, there are the special fears of the British and French that their efforts to preserve national strategic nuclear forces would be put under severe strain by the development of a defence network surrounding the Soviet Union.

By definition, the second set of European interests come into play as soon as there is any evidence that the super-powers are embarking upon an ABM race, while the first only become relevant when, if ever, there are signs of one side reaching the finishing post or even of a pending dead-heat. Moreover, the gap between embarkation on an ABM programme and it yielding results is likely to be lengthy. This gap will ensure that European attitudes will be governed for many years by their immediate reactions: probably that the whole ABM effort is unsettling and therefore undesirable. It will take time before Europeans will seriously consider the consequences of such a decisive shift in the strategic balance. Even then, their reactions are still likely to be unfavourable.

This negative assessment will be a great disappointment to those in the US defence community who believe that a successful ABM programme would inject new credibility into the US strategic position. However, this disappointment is unlikely to influence US decisions. A reading of the current American interest in ABMs suggests it is driven by the search for some technical fix to the problem of ICBM vulnerability rather than as a political fix to the problem of general European vulnerability. This search to solve the problem of ICBM vulnerability has been a constant feature of US strategic debate since 1969 and is now getting somewhat desperate, with the intrusion of political and financial reality into the consideration of basing modes of the MX missile.

There are no signs that European views are believed to be at all relevant on this matter. For example, the report of a conference held in

Washington in Spring 1981 on the ABM issue finds not one contributor even starting to put the issue in an alliance context. The only mention of a possible European interest is a reminder by Jack Ruina that 'US nuclear allies are not going to be very enthusiastic about ABM deployment, given their uneasiness about the adequacy of their own nuclear deterrent'. There is not a mention of the political impression that such a deployment might create in Europe.[1]

The risk is therefore that such a vital decision could be taken in the belief that it is a matter of prudent technical adjustment, concerned solely with ensuring the survivability of US retaliatory forces, while ignoring the effect on a European public opinion that is already in quite a state on nuclear issues. To illustrate the risk it will be useful to examine the first great ABM debate of 1966-1972.

Europe and the First ABM Debate

Anti-ballistic missiles have not traditionally been taken as an alliance issue. Symptomatic of this was the American announcement of September 1967 that a 'thin' defence, known as the Sentinel system, was to be deployed against some eventual Chinese ICBM threat. This was made before a meeting of NATO's Nuclear Planning Group. There was no consultation on the matter, and European views appear to have had no relevance to the decision-making in Washington.[2]

The West European governments' reaction to the Sentinel decision was negative. If taken at face value it reflected another example of America's obsession with China, with an exaggerated view of China's nuclear potential. If, as many suspected, it was a ruse to ease the US into the deployment of ABMs which would soon be re-oriented against the USSR, then this could prove to be politically and militarily unsettling. There was little expectation at the time that Europe could eventually prove to be a beneficiary of a ballistic missile defence. There was some suggestion that the crude SS-4 and SS-5 intermediate range ballistic missiles (IRBMs) facing West Europe might not be as difficult to intercept as the ICBMs facing the US, but the short flight-time of the IRBMs argued against such optimism, as did awareness of the vast array of other nuclear devices that could be used against West Europe. Attacking West Germany from the East by some nuclear means or other is not one of the most demanding tasks ever to confront a military planner. Nor was there much of an inclination to commit large sums of money to a programme surrounded by so many technical doubts. Estimates at the time suggested that the deployment of a US-developed system to protect West Europe could cost as much as $12 billion.[3]

This background of a decision-making process in the United States that considered ABMs by reference to some very distinctive interests and prejudices, a formal lack of alliance consultation and an awareness that West Europe was likely to be left out of this latest stage in the strategic arms race, did not encourage an intense or particularly well-informed public debate. Such debate as did take place only really got underway in 1967. This followed the announcement by Secretary of Defense Robert McNamara in late 1966 that the USSR was constructing its own ABM system and the consequent arguments in the United States over whether to follow suit. Because the American debate was so divided, Europeans were not faced with the problem of attempting to assess a prevailing American view, but were able to choose between alternative positions.

By and large those Europeans who took an interest opted for the liberal position in this debate—that ABMs were a snare and a delusion, unable to offer real safety from attack while possibly triggering a new, unnecessary and costly stage in the arms race. There was concern that such a development would abort an embryonic East-West detente. An opportunity to put strategic relations onto a more sensible and stable footing would be lost.[4] It was thus possible for Theodore Sorenson, reporting West European views for the main 'anti' contribution to the great ABM debate of 1969, to describe, probably correctly, European reactions to the Nixon Administration's ABM programme as if they were by and large the same as those of American liberals.[5]

There were two other features of European attitudes along with this general anti-ABM disposition. The first was a lack of interest in the sort of distinctions common in the American debate. It was hard for Europeans to take seriously the American anxiety over China, so there was little respect for distinctions between anti-Soviet and anti-China systems. Outside the technical community (which in Europe at the time did not extend far beyond official circles) there was little appreciation of the difference between hard-point and area defence. The assumption was that the issue under discussion was an area defence against Soviet attack, even though this was the least likely, in practical terms, of the American options.[6]

Instead, the debate was viewed through the filter of European strategic fears. As Ian Smart observed of the 1967-9 debate:

> One persistent element in West Europe was a type of anxious irritation—not with reference to America's predicament but with reference to America's preoccupation. Outside technically sophisticated government circles, the military arguments for and against ABM deployment fell on deaf ears. What was generally recognized was the implication, justified or otherwise, that the United States might be about to insulate itself from its European allies,

practically and emotionally, by deploying a new defence of its own homeland.[7]

This concern with the spectre of a 'Fortress America' features prominently in all contemporary accounts.[8]

The sort of transformed strategic relationship that might be brought about by an effective anti-Soviet ABM system could appear promising as well as dangerous. The prospect of successful area defence was 'interpreted contradictorily as either a step toward withdrawal to Fortress America from the hazards of nuclear diplomacy or as a measure to increase the credibility of US deterrence on behalf of Europe by attempting to re-establish the pre-Sputnik era of diplomacy'.[9]

As Johan Holst observed, the view adopted was likely to depend on 'the existing images in Europe of the nature and development of the Atlantic Alliance'.[10] Basic prejudice would determine the assessment. An ABM would alter the whole balance of risks within the Alliance: the risk to America in nuclear crises would decline while that of its allies would remain the same. The inequalities in NATO would be accentuated. Anything which made it easier for the US to issue deterrent threats ought to work to the benefit of the alliance, but only presuming that Washington's new freedom of manoeuvre would be used to general advantage. If the United States adopted a high-risk foreign policy as a direct result of ABM protection, then Europe might be picked upon by a frustrated Soviet leadership unable to get at its main adversary. A further transformation of the strategic relationship, by which both super-powers achieved effective defence, would not ease the latter fear but would undermine, once more, the credibility of US deterrent threats.

However, as Holst also pointed out, such dramatic and far-reaching strategic relationships were hardly on the cards. 'Marginal trends' rather than 'major shifts in the international system' were all that was at stake. Absolute invulnerability was simply not an option, and therefore the systems under consideration would not lead to the sort of profound changes feared (or desired). The sort of ballistic missile defences under consideration would 'not constitute a sufficient or even a necessary prerequisite for the development of a super-power condominium'.[11]

What this analysis suggests is that the importance for Europeans of US policy on ABMs, and indeed any significant departures in the development of nuclear weapons, had less to do with the direct impact on the US military position and much more to do with what was taken to be revealed about underlying US interests and motives.

Europe and the next ABM Debate

If this analysis of the 1960s experience is at all valid then it has important implications for the European reaction to any revival of interest in ABMs in the 1980s. Once again it is likely to take time for the issue to attract much notice in Europe and, when and if it does, there will be the same disinclination to make distinctions between different types of ABMs. The cognoscenti believe that there is no comparison between a dedicated hard-point defence and area coverage, and so will become exasperated at the general assumption that such distinctions are unduly subtle and largely irrelevant. Even many who recognise the distinctions will argue that a hard-point defence is the 'thin edge of the wedge' and that demand for an area defence will inevitably follow.

The evidence that the Soviet Union is close to a 'breakthrough' in ballistic missile defence is circumstantial and not highly regarded in West Europe.[12] Progress in those futuristic concepts is not expected for many years, while there are few signs that the USSR has an equivalent interest at the moment to that of the US in hard-point defence. If the USSR did take the initiative to break out of the constraints of the 1972 Treaty then it would suffer the opprobrium, and a US response would be accepted. However, it is more likely that the initiative will come from the US.

Partly but not only because of this Treaty the reactions in Europe to such a US initiative may well be much stronger than in the 1960s. The reason for this is the current large-scale agitation in Europe concerning alleged US plans for 'limited nuclear war'. The claim is that US nuclear forces are being based in Europe so as to confine any future war to the continent. The East-West conflict, it is argued, will be fought out over European soil even if the immediate cause of hostilities has nothing to do with the allies and results from American actions that they oppose.[13]

This set of fears is based on the divergence in US and West European perceptions of each other's interests and behaviour. It also derives from a misinterpretation of past practice, existing and planned capabilities and trends in US strategic doctrine. A move to ABMs would exacerbate both these factors.

In terms of the analysis of US strategic doctrine it would tend to support a perception of Washington's underlying indifference to the fate of Europe. It can be argued against the protestors that cruise missiles are quite inappropriate as instruments of limited nuclear war. As American missiles that are prepared for attacks against Soviet territory, their use would be less likely to keep a nuclear war limited than to ensure that it engulfed both super-powers. But if the possibility existed of a Soviet retaliation being caught by US defences, then all the worst suspicions would be confirmed. It would hardly be of comfort to Europe to host US offensive missiles so complicating to the Soviet defensive task, while Americans rested securely behind an effective defensive barrier. The Europeans would feel terribly exposed.[14] As in the 1960s, the likelihood of the worst fears being realised is very slight. An expressed US interest in hard-point defence would not justify the spectre of 'Fortress America'; but as was argued earlier new departures in nuclear weapons tend to be understood, in the context of existing prejudices, as revelations about the basic character of the responsible power.

A move to ABMs which meant the end of the 1972 Treaty would have a direct impact on European attitudes. The anxiety over the nuclear arms race has encouraged a firm commitment to arms control in Europe. European political leaders put a lot of effort into persuading the Reagan Administration of the vital importance of being seen to be exploring diplomatic alternatives to an arms race. On 30 November 1981, US and Soviet negotiators began discussions on Intermediate Nuclear Forces, a category which includes the cruise and SS-20 missiles which have generated the greatest alarm in Europe. These talks are described as taking place 'within the SALT context' and depend on substantial parallel movement taking place at the intercontinental level. These core strategic arms talks became stalled with the US Senate's unwillingness to ratify SALT II. New negotiations are now under way with START, ostensibly designed to produce real reductions at the strategic level. It would be fair to say that the Reagan Administration has yet to convince European opinion that there is any real flexibility in its negotiating position or that it is at all serious in the pursuit of arms control.

American protestations of interest in an arms control route to security would therefore be unlikely to survive any abrogation of the ABM Treaty, particularly given the non-ratification of the SALT II Treaty. If the only substantial achievement of arms control were removed the whole enterprise could collapse. If this risk was taken in order to make possible a series of marginal adjustments to the US force posture then it would simply not be understood in Europe except in the most sinister terms. At the very least it would be taken to symbolise the high priority attached by the Reagan Administration to purely military considerations as against the broader needs of foreign policy. Unless this image of American priorities is corrected, there will be increasing questioning in Europe of the wisdom of too close a connection with the United States.

Britain and France

While the rest of Europe may assess a new 'ABM race' in these broad political terms, Britain and France must consider its effects on their own military planning. The problem is not new, for the Soviet deployments of ABMs in the 1960s raised severe doubts about the British and French ability to remain viable nuclear powers. Their position was complicated by the apparent feasibility of a thin ABM protection against small nuclear powers.

The two countries were unimpressed by American claims in the 1960s that ABMs could constitute an important anti-proliferation measure, as this measure could well cut them out of the business along with more reckless would-be nuclear powers. American policy at the time was still hostile to the development of extra nuclear capabilities in NATO, and this had been a major cause of the friction with France (Britain, as a long-established nuclear power, received more favourable treatment).

The fact that the Soviet ABM system might negate the European nuclear forces was one of the few things that might have commended it to Washington. McNamara's concern was with the bad impression a US anti-Soviet ABM might make on Moscow. He seemed to have been less aware of the adverse impression in Europe made by the anti-China Sentinel system, which was seen as a symptom of excessive Sinophobia and, in being directed against a small nuclear power, as an unfortunate precedent.

The possibility that an effective Soviet ABM system would force Britain and France out of the nuclear business was widely recognised at the time. In Britain, for example, the irony was pointed out of such a strong Soviet counter-move just as the first Polaris missiles were being prepared for initial operational deployment.[15] However there is some evidence that officials in both British and French governments were more sanguine about the actual problems likely to be caused by Soviet ABMs.[16]

The British response to the Soviet 'Galosh' system surrounding Moscow is now reasonably well documented. The announcement by Secretary of Defense McNamara in late 1966 that the US intelligence community was now convinced that 'Galosh' was for the purposes of ballistic missile defence[17] served as the cue for Britain's small nuclear élite to consider the implications. Britain could have opted for the same response as the United States—a move to MIRVing with the Poseidon SLBM. Politically this was not really an option for a Labour Government that had been able to persevere with Polaris only because the costs had not proved to be excessive and because there had been no need to draw attention to the nuclear force. Those officials who would have liked to stay in step with the United States recognised that there was no immediate need for Poseidon.[18]

However, the division of the Poseidon front-end into numerous separate warheads was designed to accomplish two functions, only one of which was of interest to Britain. The early separation of the warheads, coupled with other penetration aids, served to complicate enormously the defender's task. The ability to aim the individual warheads independently served to increase the potential target structure. As Britain had only reasonably simple targeting requirements, prime interest was in the first of these features.

In 1967 British nuclear scientists began a research project geared to designing a warhead that concentrated on penetrating ABMs, even if this emphasis came at the expense of the number and variety of individual targets that could be attacked. The research programme, initially drawing on US concepts developed under the Antelope project of the early 1960s, looked particularly at warhead hardening to protect against the effects of exoatmospheric nuclear explosions from Soviet ABMs, and decoys to confuse Soviet defenders.

In 1972 the Conservative Government considered whether to push forward with this indigenous project or to purchase Poseidon from the US. It decided against Poseidon, largely on the grounds that while an anti-ABM system might be required, a full MIRV system would not be appropriate for Britain. By the early 1970s MIRVing had started to acquire the same bad name in Europe as had ABMs a few years before. Politically life would be easier without having to justify acquiring MIRVs in another highly public deal with the US. Persevering with a secret national programme, with a much less controversial objective in mind, would be easier. It was recognised, probably correctly, that a future Labour Government would be unable to endorse Poseidon. This general political judgement was supported by technical and cost estimates, which favoured the British programme. These later turned out to have been extremely optimistic.

The British programme, which became known as Chevaline, was supported by successive Labour and Conservative governments. Although technically it was a success, in providing a manoeuvering, hardened, decoy-based front-end for Polaris, this success came at greater cost (£1,000 million) than expected and its introduction was delayed. It was not scheduled for introduction into UK forces until 1982.[19] By this time, and partly because of the experience with Chevaline, the British Government had decided to adopt MIRV technology along with the US Trident missiles for the replacement of Polaris in the early 1990s.

For the purposes of the current essay it is important to note that the decision to proceed with Chevaline was taken *after* the 1972 ABM Treaty, when the USSR became restricted to 200 launchers. There were only 64 launchers in the 'Galosh' system at the time, and by 1974 it should have been clear that there were unlikely to be many additions.[20] The rationale for Chevaline was that if Britain was to have a credible deterrent then it was necessary to be able to threaten Moscow.[21] Chevaline was designed to meet the specific problem posed by the 'Galosh' system. Under the 1972 Treaty the quantity of ABMs if not the quality was restricted, and this made it a manageable problem. If, as currently planned, Britain introduces the Trident system in the early 1990s (either in the C-4 or D-5 variant) then it is likely to be in a position to penetrate any conceivable defences, at least for the rest of this century.

France, too, has designed its forces so as to be able to operate in the post-1972 ABM environment, though the issue has not been as fully discussed as in Britain. It was reported that in the 1975 series of tests, the hardening of the warheads was a major concern.[22] The M-4 warhead under development for introduction in 1985 with the submarine-based force, is to be used in a six warheads MRV system. Although this is not a MIRV system, so the warheads are not independently targeted, it may well be that they separate early enough to pose major problems for the defending forces.

It is commonly believed that both the British and French forces are vulnerable to any future enhancement of ABM capabilities, either in an expansion of defensive systems following abrogation of the 1972 Treaty or in the development of some of the more exotic new systems, such as those based on directed-energy. One critic of the British government's choice of Trident argues that the pending arrival of laser or particle-beam defences is likely to undermine deterrent based or ballistic missiles.[23] This view has naturally enough been rejected by the Government.[24] Leading British defence scientists are privately quite sceptical as to the likelihood of any breakthrough in the field of directed-energy weapons being translated into a viable defensive system until well into the next century if at all. Nevertheless, even if this more exotic threat is discounted, the British and French governments have made no secret of their anxiety that the 1972 ABM Treaty should not collapse.

Yet even here the risk to the viability of the European deterrents may not be as great as is often supposed. If the United States does decide that some form of hard-point defence is absolutely essential to the survivability of at least a portion of its ICBM force, and that the allowance under the 1972 Treaty (as amended in 1974) of 100 launchers is insufficient, then it is most likely to approach the USSR to amend the Treaty to make possible hard-point defence but to continue to exclude area defence. It is unlikely that the Soviet Union would agree to such amendment, but it is also unlikely that if American pressure led to the collapse of the Treaty by other means, the USSR would move to a massive new investment in area defence, given the advantages that still accrue to the offence.

Even if the USSR was only interested in defending itself against Britain, France and China that would still require a major investment. It would not be enough to spread large numbers of ABMs around Soviet territory. On any given line of attack, many of these ABMs would be useless. As the mobility of the British and French submarine allows for attacks to be launched from a great variety of points, there would be many lines of attack to be covered. To accomplish this a massive network would be required. It might well be possible for the USSR to protect Moscow effectively, by enhancing substantially existing capabilities, but it would be difficult to deny the smaller nuclear powers access to all economic and political centres.

So as a practical matter, the end of the ABM Treaty need not mean the end of the British and French deterrents. But as a political matter enormous public doubts about the viability of the national strike forces would almost certainly be raised. Officials would find it difficult to insist that little had changed and that no counter-measures were necessary. And if counter-measures were deemed necessary then that would raise the whole question of cost. Particularly in Britain, where there is already a widely-held view that even a minimum deterrent is an unwarranted burden on the public purse, any development which made it difficult to be confident in existing or planned force levels could prove to be decisive in the continuing argument over the value of retaining a national nuclear force.

Conclusion

This analysis has stressed the importance of prejudice and perceptions in shaping European attitudes to anti-ballistic missiles. Given the current state of West European public opinion on the nuclear issue, with the widespread anxiety that the nuclear arms race is once more getting out of control, and in such a way as to create novel dangers for Europe, a stress on defensive technologies would just add to the prevailing sense of danger and anxiety.

Whatever the ultimate potential of the new defensive technologies, Europe would take some convincing that the opportunity to re-order the strategic relationship based on something other than mutual super-power vulnerability could be of benefit to them. If the super-powers could defend themselves effectively then the European vulnerabilities would be accentuated, for not only could West Europe not be protected but also the British and French nuclear forces would appear ineffectual.[25] Even if none of these dramatic developments ever came about, the very act of making an effort to promote the relevant systems would prompt suspicions in Europe.

Alternatively it might be claimed that the US intention was to achieve no more than an added degree of survivability for the land-based ICBM force, and that though it might entail altering the ABM Treaty this would add to, not detract from, the stability of the strategic balance. Even if the Europeans could be convinced that US objectives would stay so modest, they would be unlikely to understand why it was worth jeopardising the whole arms control process for such a minor improvement. Either way the severity of the political reactions to a new ABM race would probably far outweigh the actual shifts in military relationships that could eventually be achieved.

NOTES

1. Conference on 'ABM Revisited', *The Washington Quarterly*, VI:4 (Autumn 1981). The Ruina quote is from p. 65. Consideration to the alliance dimension is given by Albert Carnesale, 'Reviving the ABM Debate', *Arms Control Today*, II:4 (April 1981). He is certain of the negative allied reaction to Soviet BMD deployment but less sure of the reaction to US deployment.
2. As negative evidence one can note the total absence of any mention of alliance considerations in Morton Halperin, 'The Decision to Deploy the ABM', *World Politics*, XXV (October 1972).
3. *Economist,* 3 Dec., 1966.
4. These impressions can be gleaned from the two most informed analyses of West European views: Laurence Martin, 'Ballistic Missile Defense and Europe', in Eugene Rabinowitch and Ruth Adams (eds.), *Debate the Antiballistic Missile* (Chicago: Bulletin of the Atomic Scientists, 1967); Johan Holst, 'Missile Defense: Implications for Europe', in Johan Holst and William Schneider Jr. (eds.). *Why ABM? Policy Issues in the Missile Defense Controversy* (New York: Pergamon, 1969).
5. Theodore C. Sorenson, 'The ABM and Western Europe' in Abram Chayes and Jerome B. Wiesner (eds.), *ABM: An Evaluation of the Decision to Deploy an Antiballistic Missile System* (New York: New American Library, 1969).
6. The Sentinel programme was for area defence against Chinese attack; the Safeguard programme (of the Nixon Administration in 1969) was to protect ICBM silos against Soviet attack.
7. Ian Smart, 'Perspectives from Europe', in Mason Willrich and John B. Rhinelander (eds.), *SALT: The Moscow Agreements and Beyond* (London: Collier Macmillan, 1974), p. 187.
8. Thus Sorensen suggested that whatever the declared role of the Safeguard system in protecting ICBMs, it would be seen in a more sinister light. '(M)any if not most West Europeans will believe instead that the United States is increasing its capacity to ignore some future Soviet nuclear threat which European nations cannot escape.' *Op. cit.,* p. 179.
9. Martin, *op. cit.,* p. 125.
10. Holst, *op. cit.,* p. 196.
11. *Ibid.,* p. 190.
12. One author that does accept claims of Soviet progress in directed-energy technologies is Dr. David Baker, *The Shape of Wars to Come* (London: Patrick Stephens Ltd., 1981). This book opens with a 'scenario' for 1995 in which the USSR introduces a 'ground-based particle beam device designed to screen Soviet territory from incoming warheads', while the US is rushing to get in place Star Raker, a space-based anti-missile beam weapon. The result is 'stalemate: nobody can strike the other with the holocaust of armageddon' (pp. 8-9). Unfortunately for this scenario, the position of Western Europe as Soviet hostages in such circumstances is not discussed. See also Air-Vice Marshal Stewart Menaul, *Countdown Britain's Strategic Nuclear Forces* (London: Robert Hale, 1980).
13. For a discussion of the arguments behind the current movements see Lawrence Freedman 'Limited War and Unlimited Protest', *Orbis* (Spring 1982).
14. At a conference in Washington just after the 1980 Presidential election, I was witness to a prominent Reagan supporter (an academic who later was appointed to a job on the National Security Council staff) doing his best to promote such fears. After stressing the importance of NATO's long-range theatre nuclear force modernisation

programme being implemented, he then mused aloud about how the US had fallen so hopelessly behind in the ICBM race that it might be sensible to concentrate on defensive technology.

15. For example the *Economist*: 'On the assumption the anti-missile technology is now really getting under way, the only safe estimate is that by the mid-1970s Britain's Polaris missiles will be capable of inflicting only marginal damage on the Russians' (28 Oct., 1967).

16. This seems to be the message of Martin, *op. cit.,* p. 126.

17. There was also intense debate over whether a series of installations known as the 'Tallin Line' were also for ABM purposes. It later transpired that they were for air defence purposes. For the background to the US intelligence debate see Lawrence Freedman, *US Intelligence and the Soviet Strategic Threat* (London: Macmillan, 1977), Chapter Five.

18. Much of this section is based on the relevant chapters in Lawrence Freedman, *Britain and Nuclear Weapons* (London: Macmillan, 1980).

19. *Times,* 30 Jan., 1982.

20. A 1974 Amendment brought the permitted numbers down to 100. In 1980 the USSR cut the number of launchers in the 'Galosh' complex from 64 to 32, probably in order to bring in a replacement. The possibility was, however, raised that Chevaline might now be a battering ram for an open door.

21. Whether or not this makes much sense as strategic theory is not relevant to this article. For a documentation of the theory see Freedman, *Britain and Nuclear Weapons,* Chapter Five.

22. *Times,* 10 June, 1975.

23. See Menaul, *op. cit.,* and Letter to *Times,* 5 July, 1981. Curiously Menaul appears to believe that cruise missiles will still be able to survive the breakthroughs in defensive technology, though most analysts would argue that cruise missiles are already vulnerable to known means of defence.

24. Defence Secretary John Nott when asked about Menaul's letter (see note 23) said that there were no grounds for believing that a system would be developed to knock out ballistic missiles in the next decade. *Times,* 9 July, 1981.

25. For an interesting discussion of the possibilities of damage-limiting anti-tactical ballistic (or cruise) missiles for Europe see David Yost, 'Ballistic Missile Defense and the Atlantic Alliance', *International Security,* Fall 1982 (Vol. 7, No. 2). Despite his own sympathy for such a project the author is unable to find much evidence to suggest that this sympathy would be widely shared in Europe.

Extended Deterrence and the Protection of the United States ICBM Force

Ian Bellany

There are two ways of protecting United States' intercontinental ballistic missiles (ICBMs) from Soviet attack, just as there are, for NATO, two ways of protecting Western Europe from Soviet attack. One way is through defence; another way is through deterrence, or, more precisely, extended deterrence.

The notion that the protection of Western Europe depends both upon defence and extended deterrence is a familiar one: that extended deterrence may also have a part to play in the protection of the United States' ICBM force will be less familiar, but it is the central proposition of this essay.

Questions of extended deterrence normally arise whenever consideration is given as to how a nuclear power may make use of an undoubted capability to devastate the greater part of its opponent's cities and wealth producing centres, even after absorbing a determined attack on its capability so to devastate, to deter not only an attack on its own cities and wealth producing centres, but also attacks upon targets of lesser or secondary importance, as perceived by the other side.

A useful rule of thumb for gauging the probable effectiveness of extending deterrence in any given situation is due to Rosecrance.[1] It stresses the significance of relative capability on the one hand, and the degree of importance, or centrality, of the target to be protected on the other, as being jointly influential in determining the effectiveness of the extended deterrence.

Thus when the United States enjoyed a substantial margin of nuclear superiority over the Soviet Union—a large, favourable relative capability in other words—extending deterrence to protect Western Europe from Soviet attack, whilst never exactly a simple matter, was comparatively straightforward. And to give a contrasting example, the current position of the United Kingdom of substantial nuclear inferiority with respect to the Soviet Union means that the United Kingdom cannot be in a strong position to extend nuclear guarantees to allies, but it may be well enough placed to deter a Soviet attack aimed at the destruction of London and the other chief cities of the country, since the centrality of these targets can hardly be in doubt.

Taking the protection of Western Europe as a model for the protection of the United States' ICBM force, there are good historical reasons why the role of extended deterrence has always been prominent enough in the former case yet all but ignored in the latter. When the United States first accepted a commitment to Western European security, the very possibility of effective defence was in question and deterrence came accordingly to the fore. Latterly, with the complete recovery of the Western European economies and the direction in which most technical developments in the design of conventional weapons seem to be heading, the possibility of defence is no longer doubted and deterrence has retreated. In the case of ICBMs, except for a short period when they were first developed which is alluded to briefly below, defence was not originally a worry—silo basing was cheap and effective and was so to remain for almost two decades. But with the coming into general service of the multiple targetable reentry vehicle (MIRV)—in effect a highly accurate and reliable means of delivering very large numbers of warheads over intercontinental ranges—the defence of the ICBM has now begun to pose new problems. The casting about by the present United States' government, and its predecessor, for a safe way to base the planned successor to Minuteman, the MX ICBM, and the picking up and putting down of one basing scheme after another are ample illustration of the difficulties involved.

If the trend of technology is indeed such as to place the defence at an increasing disadvantage, but the United States remains unwilling to abandon the ICBM part of its so-called triad of strategic nuclear forces, the answer may lie in extended deterrence. And obviously for extended deterrence to be of any use in an era of parity of capability it will be essential to play skilfully upon Soviet perceptions of what the consequences would be of an attack on the United States' ICBM force.

Of course were the Soviets of the opinion that the survival of the Minuteman ICBM force was a matter of supreme importance to the United States the battle would be more than half won. But, ironically, one of the things in the way of the Soviets coming to a conclusion of this kind is the existence of the United States' triad. If Minuteman were literally the *only* means the United States possessed of delivering nuclear blows on the Soviet Union, its importance to the United States would be self-evident. The Soviets would have to ponder whether an attack on Minuteman would not automatically lead to a desperate but fearsome retaliation, possibly using a mixture of launch on warning and launch under attack, on Soviet cities. As things stand, the existence in considerable strength of United States' submarine launched ballistic missiles, bomber forces, and forward based delivery systems in the European theatre signals something quite different: namely that the loss of Minuteman could in some circumstances be tolerated.

The same reasoning plainly does not apply to the Soviet ICBM force. With no forward based nuclear delivery vehicles, a small and obsolete true intercontinental bomber force, and a fleet of ballistic missile submarines of limited dependability for second strike purposes, the ICBM is the backbone of the Soviet nuclear capability. The very dependence of the Soviets on their ICBMs provides (via extended deterrence) appreciable protection against their being attacked.

Nonetheless extended deterrence may still have something to offer the United States. Renewed emphasis on the importance accorded to the ICBM part of the triad, in whatever form, even by arranging for the existing mode of fixed site bases to receive ABM coverage (whilst not a very cost effective mode of defence) would make some psychological sense. It would signal to the Soviets the importance the United States attached to the survival of its ICBM force and its willingness to make considerable sacrifices—in this instance of a financial, or opportunity cost kind—to do something about protecting it. Seen in this light, even the costly and cumbersome proportions of the now cancelled multiple protective shelter (MPS) scheme for the MX ICBM were a virtue.[2] The impression that the United States could not do without its ICBM force would be further strengthened if militarily valuable features, such as high accuracy and quick retargeting capability were allowed to remain an ICBM monopoly, even if and especially if, the technology existed to permit these features to be incorporated in submarine launched missiles and elsewhere.

Alternatively, or, with a little ingenuity, additionally, the United States could base its ICBMs in such a way as to deny itself an easy decision that a Soviet attack on its ICBMs was something not meriting an American retaliation on Soviet cities. Already, depending upon weather conditions, the fission megatonnage used in the attack, the time of day and the amount of warning available to the civilian population, it is generally acknowledged that a Soviet attack on the existing Minuteman silos could lead to the subsequent death from fall out radiation of several million of those American civilians that live down wind of the Minuteman wings. The various imponderables mentioned, and the time lag between the attack and its effects on the civilian population, make this an unreliable and limited mechanism for extending deterrence, but the same principle can be applied more subtly.

In the 1950s, for a brief period when ICBMs were new and, because of the type of rocket fuel then employed, very slow to bring up to, and take down from, a state of readiness for launching, they were considered highly vulnerable to surprise attack, and the impractical and only half serious suggestion was made that they should be based in the hearts of American cities, where the Soviets would not dare to attack them because they would know what kind of retaliation to expect.

An up-dated and more practical version of the same idea can be employed to make a virtue out of what is usually seen as a drawback of the air mobile basing mode that was a widely canvassed rival to MPS/deceptive basing for the planned MX missile.

The supposed drawback is that while the launch aircraft carrying the ICBMs are essentially safe from attack once airborne, the airfields they need to land on for refuelling and servicing (unless a capability for retaliation within the few hours' endurance of the launch aircraft was all that was required) are vulnerable, and it would be impossibly costly to build more of these recovery airfields than the Soviets could obtain warheads to hit them with. Yet an operational procedure that allowed, encouraged and made provision for the launch aircraft to make emergency use of the *commercial* airports of American cities would force the Soviets into the extremely awkward position of having to attack these cities, or appear to do so, if they wished to deny the air mobile ICBM recovery sites.

It is possible, although it will not be attempted here, to make a point by point listing of the methods employed to make extended deterrence work in the NATO context in order to see how the conceptual equivalent in each case might be employed to protect the ICBM—the above suggestion as to how an air mobile system ought to be organised has an obvious affinity to the 'hostage' function of United States' servicemen stationed in Europe, for example. Nothing so elaborate however is required to discover that in one very important respect extending deterrence to protect the United States' ICBM force is easier than extending deterrence to protect Western Europe. Even when the nuclear superiority of the United States over the Soviet Union was at its height, extended deterrence could never deal with every conceivable Soviet 'threat'. In particular a Soviet adoption of 'salami' tactics was foreseen as a likely problem—extended deterrence could hardly be expected to cope with a Soviet use of conventional force from time to time to adjust an international boundary in their favour by a few miles. And one of the soundest of the original justifications for a NATO conventional capability was to push upwards the amount of force the Soviets would need to exert in order to secure even a small objective, and so increase the chances of extended deterrence proving effective. No such problem arises in extending deterrence to protect the United States' ICBM force. The natural response (and the natural deterrent) to a small attack on a few ICBM bases is a similar counter-attack on Soviet ICBM bases.

Finally, it has not been the intention of this short essay particularly to advocate extended deterrence as a means of keeping the United States' ICBM force safe, although it is not unsympathetic to the idea at least as an interim measure. Rather it has been to serve as a reminder that extended deterrence is there. Whether or not the United States decides explicitly to make use of extended deterrence, American actions and public statements with respect to its ICBM force (and the rest of the triad) are always liable to be read by the Soviets as signals whose effect may be to add to, or more crucially subtract from, the degree of security the ICBM force already implicitly derives from extended deterrence.

NOTES

1. Richard Rosecrance, 'Strategic Deterrence Reconsidered' in *Strategic Deterrence in a Changing Environment*, edited by Christopher Bertram, Gower and Allanheld, Osmun and Co., Farnborough and Montclair New Jersey, 1981, pp. 7, 31. His rule of thumb in fact speaks of a multiplicative relationship between relative capability and the centrality of what is to be protected as determining the effectiveness of extending deterrence.
2. If the 'closely-spaced basing' or 'dense pack' proposal is adopted as the basing mode for the MX missile it is to be hoped that this will be because of the exceptional defensive properties of the system, since its 'quick fix' image will do nothing for extended deterrence.